The Campus History Series

Case Western Reserve University

Squire Valleevue and Valley Ridge Farms

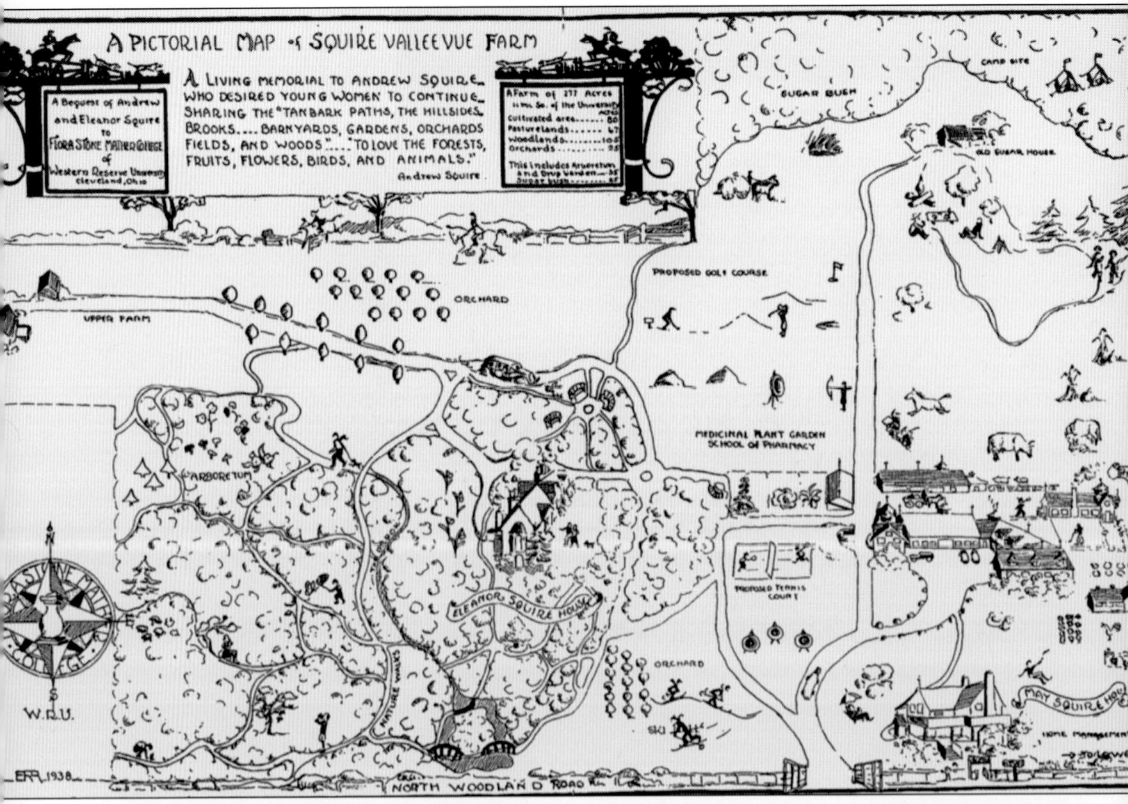

Caricature Map of Squire Valleevue Farm. In 1938, local artist, writer, and Flora Stone Mather College student Emily R. Andrews drew this map of the farm as it looked the year that it was transferred to Flora Stone Mather College. Some of the areas featured were proposed but never constructed, such as the golf course. (Case Western Reserve University Archives.)

On the Cover: Young Ladies File into Wagon. This car was purchased by the women of Flora Stone Mather College with money collected at a fundraising dance held at the Pink Pig at Squire Valleevue Farm. It allowed them to easily make the 10-mile trip between the main campus in downtown Cleveland and their country campus in Hunting Valley. (Case Western Reserve University Archives.)

The Campus History Series

Case Western Reserve University

Squire Valleevue and Valley Ridge Farms

Ana B. Locci, PhD, and Chris Bond

Arcadia Publishing

ISBN 978-0-7385-8275-7

Published by Arcadia Publishing
Charleston, South Carolina

Printed in the United States of America

Library of Congress Catalog Card Number: 2011933085

For all general information, please contact Arcadia Publishing:
Telephone 843-853-2070
Fax 843-853-0044
E-mail sales@arcadiapublishing.com
For customer service and orders:
Toll-Free 1-888-313-2665

Visit us on the Internet at www.arcadiapublishing.com

This book is dedicated to all of the students who have walked these grounds and learned in these fields, fulfilling Andrew Squire's wish that students from different backgrounds would have a place to interact with the elements of nature and the facilities in which to study and research them. It is also dedicated to this college's future students.

Contents

ACKNOWLEDGMENTS

The authors would like to thank all of the many people who were involved in making this book a reality: the Locci family—Ivan, Natty, Eddie, and Ricky; Zoey and Truman Bond; Raymond and Mary Ann Leonard Zukowski, whose generosity to the farm and this book will forever be appreciated; Frieda Wallace, Kathe Martin, and Walter Walko Jr. for the use of their family images (all images from them are cited as Wallace family); Elizabeth Wade-Sedgwick for the use of her family's images; Dr. Kenneth Kutina for his editing assistance and fact-checking; the Flora Stone Mather Alumnae Association for its continued support of University Farm; Garden Club of Cleveland member and Hunting Valley resident Dr. Mary-Scott Cebul for her help in obtaining the historical drawings for Valley Ridge Farm from the University of Iowa at Ames; Dale Rene for her help compiling and organizing the assorted and miscellaneous historical documents; Stephen and Joyce Rajki for their help in obtaining many of the original blueprints and source documents; Helen Conger and Jill Tatem at the Case Western Reserve University Archives; Alan Alldridge for his help in scanning the Elizabeth Wade-Sedgwick photographs; Christine Leary for her editing help; and all of the contributions, big and small, of the staff at University Farm, the division of student affairs, and anyone else for whom the limits of space preclude naming individually. We thank you all.

Unless otherwise noted, images are from the Case Western Reserve University Archives or the collection of Case Western Reserve University Farm.

Photographs from the collection of the Western Reserve Historical Society are abbreviated as WRHS.

Photographs and drawings from the collection of the University of Iowa at Ames are abbreviated as UIA.

INTRODUCTION

Tucked into the bucolic hills amidst the verdant horse country of Hunting Valley, Ohio, is the 389-acre treasure and Cleveland asset Case Western Reserve University Farm. Its invaluable property and many buildings were the gifts of generous donors who believed in both preserving and using the land.

The lion's share of the farm, donated in the early 20th century, was the pride of famed Cleveland attorney Andrew Squire and his second wife, Eleanor Seymour Sea. It was here that he engaged in his passion for experimenting with nature, devising vast gardens and fields that included a 900-species arboretum featuring at least one of every native species. His medicinal gardens provided herbs and tinctures for area hospitals, and the eggs and vegetables he produced were supplied to many of Cleveland's most exclusive social clubs.

Squire lost his only daughter, May, when she was a young girl. It is widely believed that the 1937 bequest of his entire estate to the women's college of Western Reserve University was in her honor, allowing others to know the beauty and bounty of nature.

Another large portion of the farm was donated by the heirs of Jeptha Homer Wade II, grandson and namesake of J.H. Wade, founder of Western Union Telegraph. Jeptha made his summer estate here at his beloved Valley Ridge Farm. The younger Wade spared no expense in producing his grand manor, enlisting the services of the famed architectural firm Hubbell and Benes and of noted American landscape architect Warren H. Manning. It was Manning who enhanced the 500-plus acres of rolling hills, dynamic valleys, glades, and groves, cutting breathtaking vistas that overlooked the nearby Chagrin River and installing hidden gardens throughout the expanse of woods. Wade employed up to 50 men to work his gentleman's farm and would consider it his paradise in the woods for the remainder of his days.

The most recent land addition to University Farm—donated in the 1990s—was two, 5-acre parcels given by the Hollister family, who, at the time, was the farm's neighbor on its northern boundary.

Today, the farm serves as an educational, research, athletic, and recreational campus. It hosts tens of thousands of students, researchers, and guests each year, engaging in dozens of diverse research activities, field and lab work, and art and nature-related classes. Untold numbers of nonstudent visitors—nature lovers, gardeners, and event guests—also stroll its grounds, experiencing the beauties of each season.

We are extremely privileged to consider ourselves stewards of this magnificent land and wish to share its splendor with all.

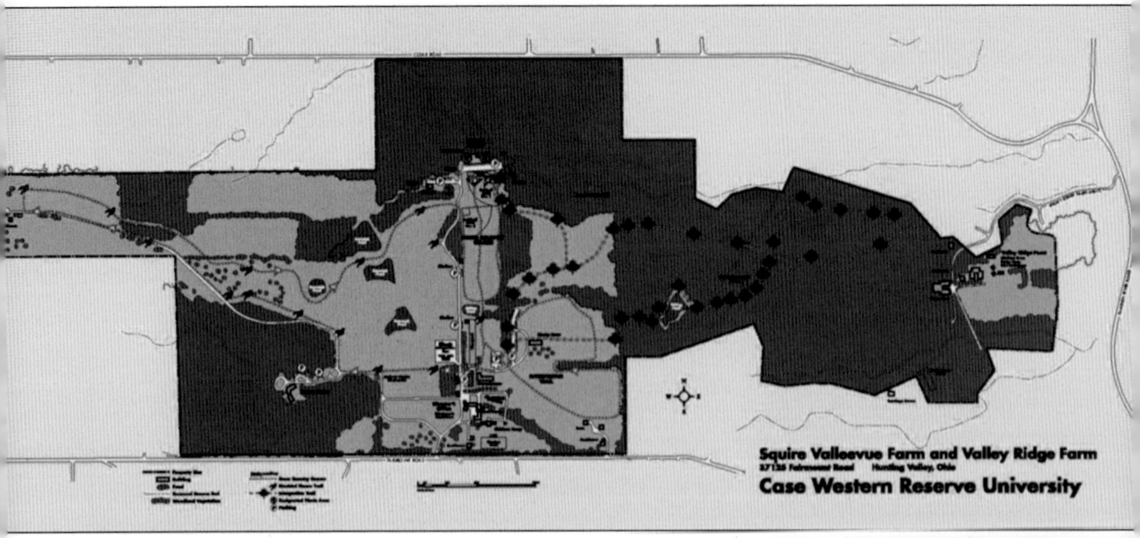

AN OVERVIEW OF THE FARM TODAY. University Farm is the combination of four gifts of land by three different families. The largest portion of its 389 acres was the gift of prominent Cleveland attorney Andrew Squire. The heirs of Jeptha Homer Wade II donated another 104 acres to the university in 1977, and two gifts totaling 10 acres were generously donated by the Hollister family, one of the farm's Cedar Avenue neighbors.

One

ANDREW SQUIRE'S FARM 1907–1937

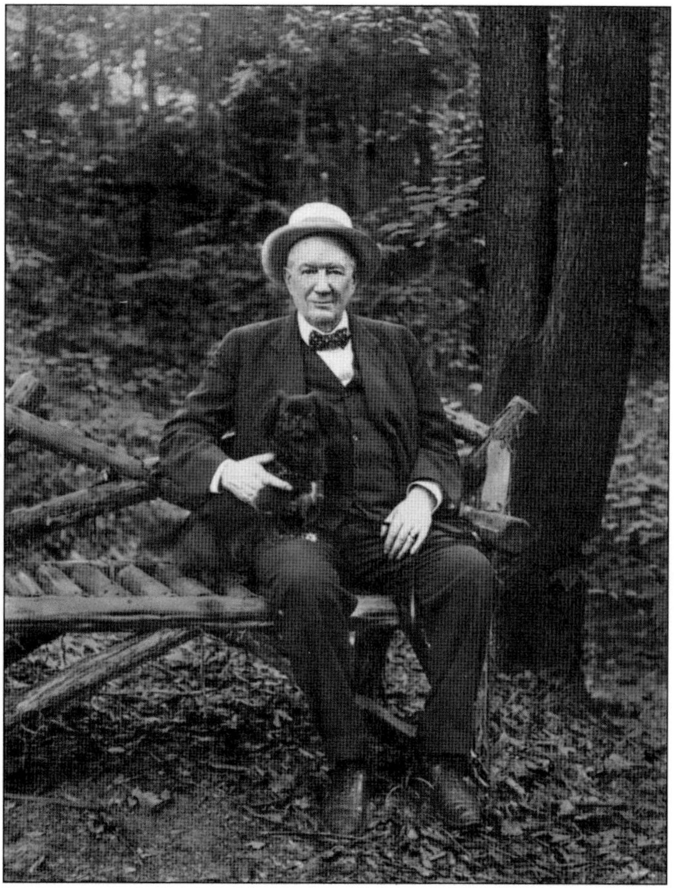

ANDREW SQUIRE (1850–1934) WITH HIS DOG. Andrew Squire (seen here, sitting in his arboretum with his beloved dog) would eventually purchase several parcels of adjoining land in picturesque Hunting Valley, Ohio, to form his vast estate, Valleevue Farm. He employed the well-known architectural firm of Walker and Weeks, of Severance Hall fame, to design his barns and buildings.

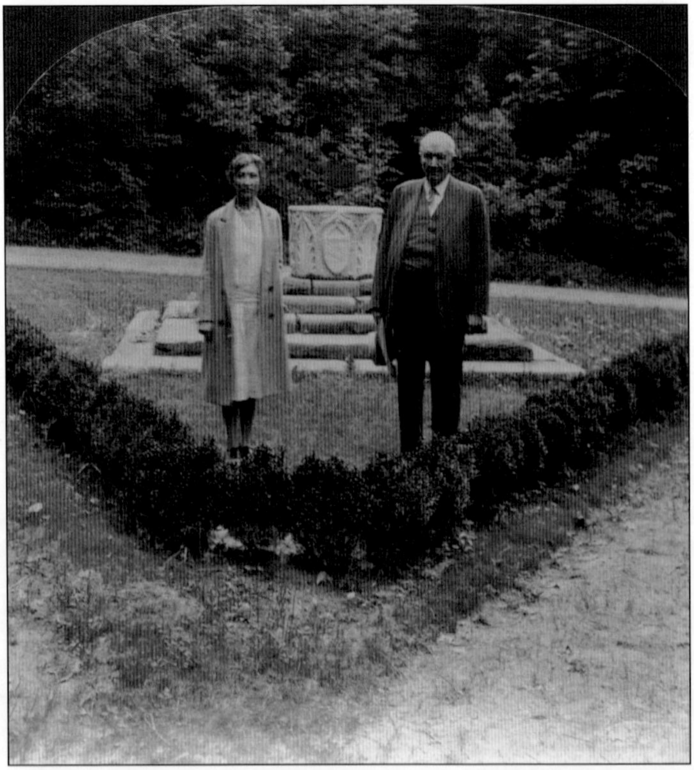

The Squires by the Wishing Well. One of the few known photographs of the couple together, Andrew and Eleanor Squire pose in front of a wishing well in their arboretum. The wishing well is still in use as one of the many decorative elements that frame the Manor House and its surrounds.

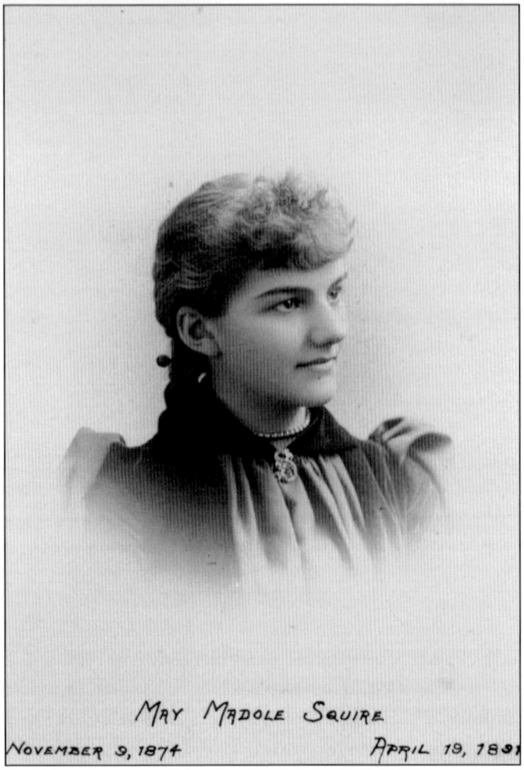

May Madole Squire. May was the only daughter of Andrew Squire and his first wife, Ellen Mott. May died suddenly and unexpectedly while away at boarding school in 1891. It was a great loss for Andrew Squire. It is believed that the impetus behind the bequeathing of his entire estate, Valleevue Farm, to the women's college of the Western Reserve University (where he served as a university trustee) was in May's honor and memory.

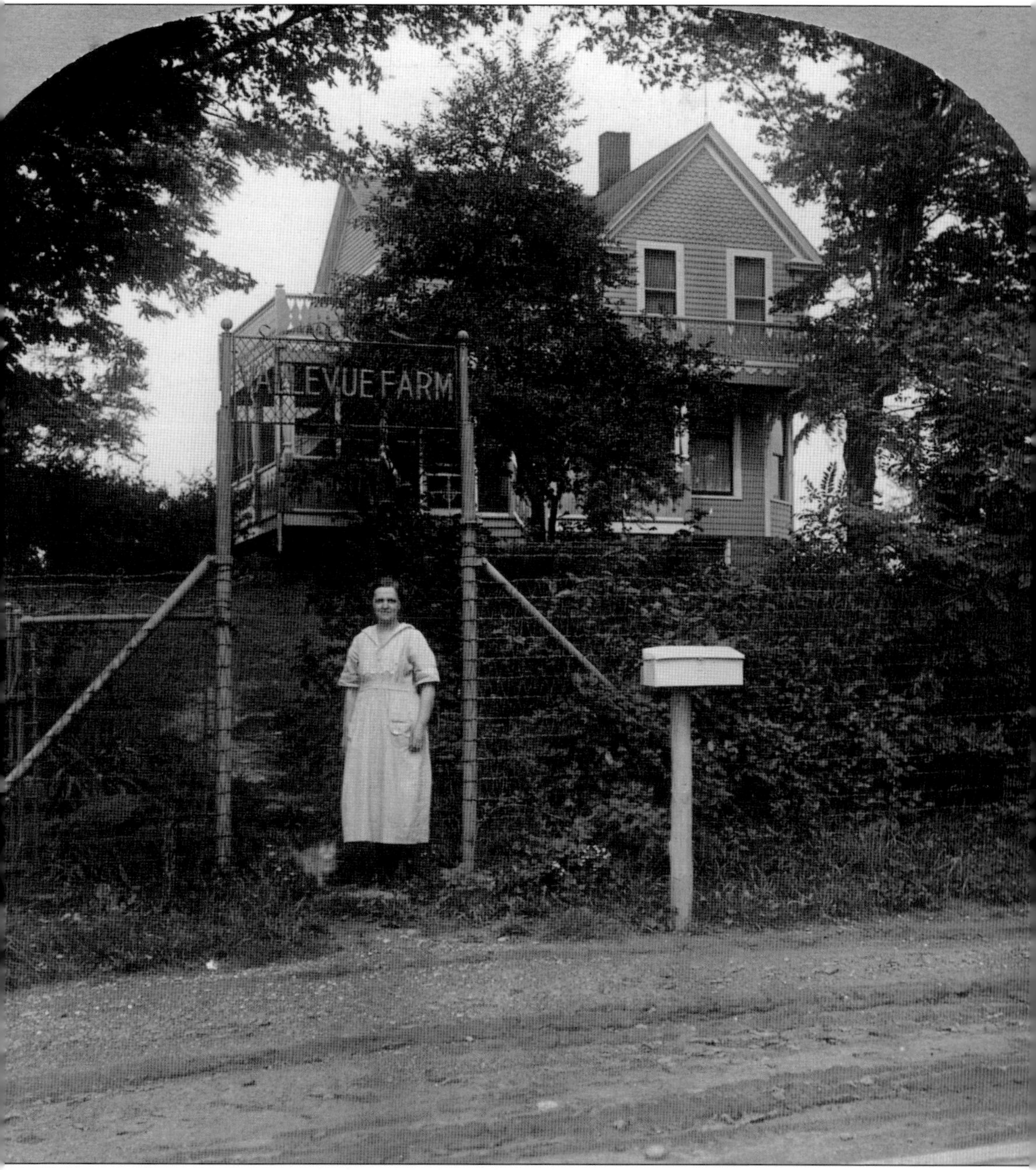

Woman by Gate (South Elevation). Squire employed up to 30 farmworkers to tend to his vast estate. The woman in this undated photograph is likely Mary Bane, whose many duties included meal preparation and laundering for all the other workers. She is standing in front of what would later become known as the May Squire House. Note the spelling of Vallevue on the gate. All other references to the property are always "Valleevue." Valle Vue or Vallee Vue is the French translation of "Valley View." If one stands at the peak of the property and enjoys the 360-degree view, it is understandable where this name arose from. The Squires were known to travel to Europe quite frequently, and their time in Paris could have been the inspiration for such a christening.

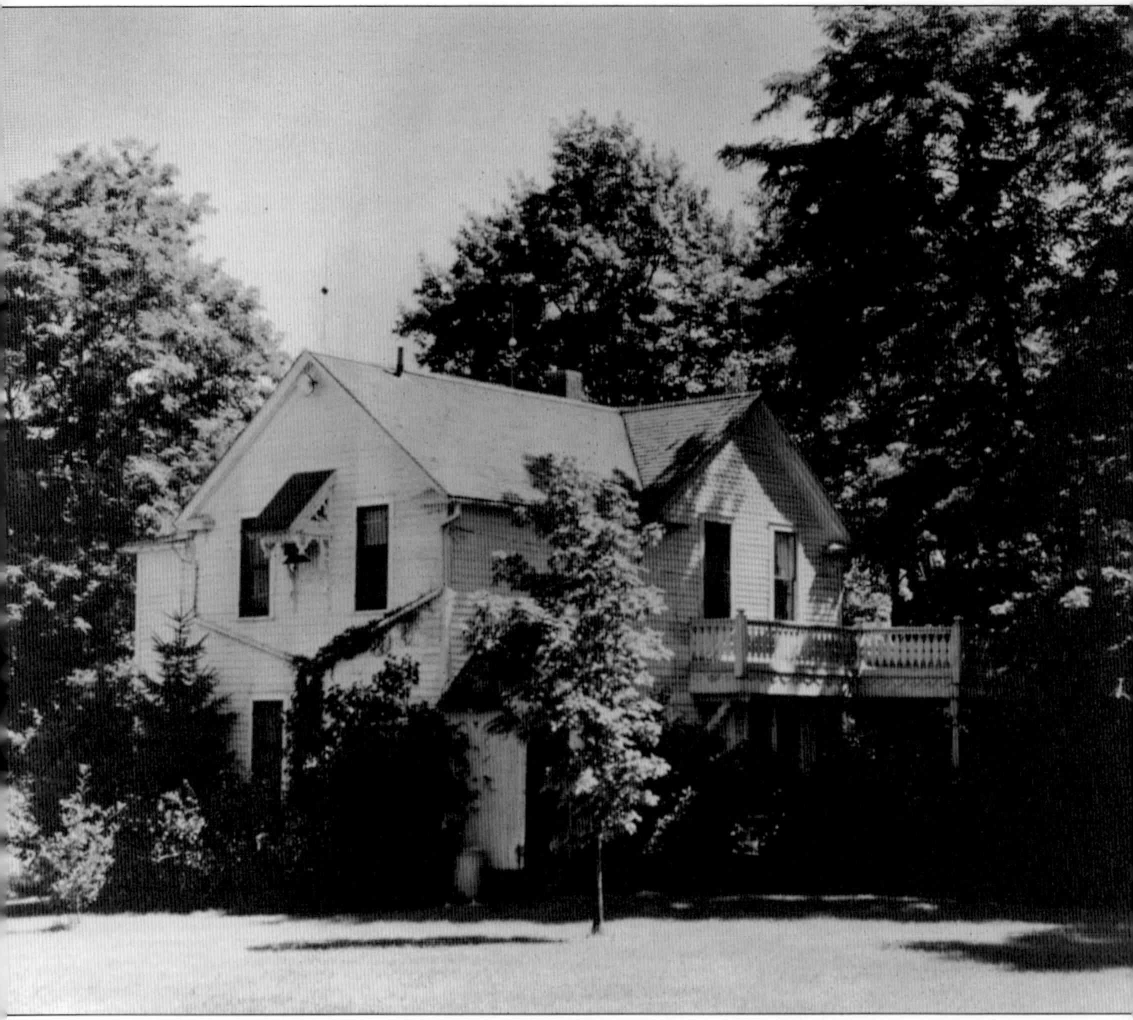

The May Squire House (Northeast Elevation). One of the original farmhouses on the property, this building is said to have been erected in 1850 by one of Hunting Valley's (then part of Orange Township) first residents, Asahel Jerome, who also started the village's first cemetery, where he is now buried. The May Squire House, so designated by the women of Flora Stone Mather College, can be seen from Fairmount Boulevard. It was first used by Andrew Squire as housing for some of his workers, such as the Bane family. It later would become and serve for over 20 years as a home economics practice classroom for the women of Flora Stone Mather College. It became the home of Dr. Norman Alldridge (the second farm director) and his family until 1976. Over the years since, it has been the residence of several university faculty and staff and their families.

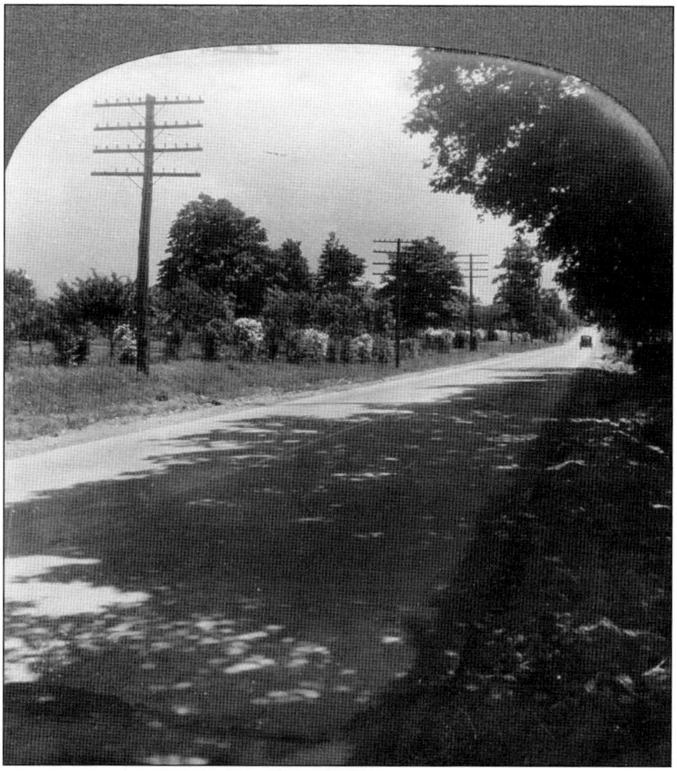

S.O.M. Center Road Entrance. This stereoscopic image, dated from 1928, shows a southerly view looking down S.O.M. (Solon, Orange, and Mayfield) Center Road in the springtime, just past the Farm's western entrance. This entrance, though it is still present to this day, no longer serves as the entry point to University Farm but is instead a driveway to one of the private residences on the farm. Flora, like that depicted here, can still be seen in abundance.

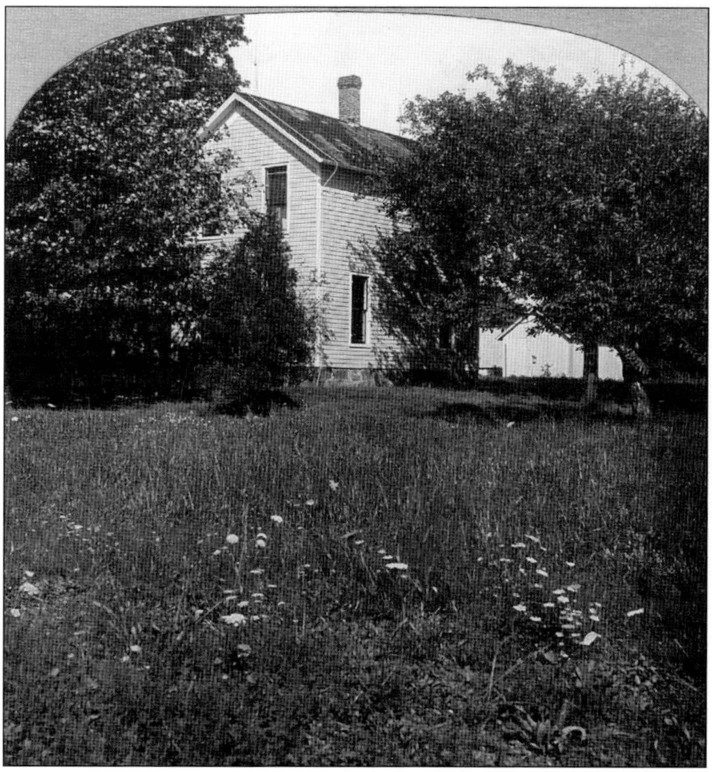

Farmhouse on S.O.M. Center Road. Andrew Squire's Valleevue Farm was not bought as the full 275 acres in one large parcel. It was the combination of three different working dairy farms dating back to the middle of the 19th century. He would start assembling his vast estate with the first parcel, which was purchased in 1907. This 1923 image shows one of the farm's residences, though somewhat obscured by the trees in the photograph.

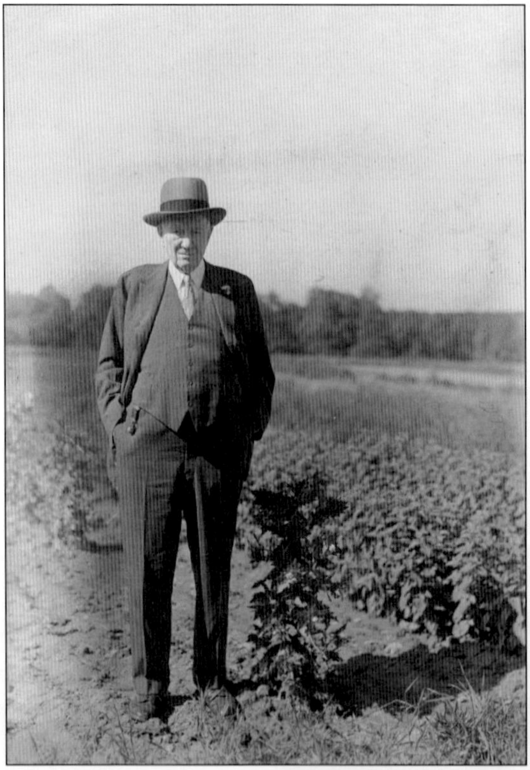

Andrew Squire Standing in His Fields. Squire (in his 70s) stands among rows of one of his many crops. Until his death at age 84, he took great joy in walking the fields, gardens, and woods on the grounds of his estate. He was known to report evidence of the first sign of each new season to his friends, such as the first spring bulbs and autumn leaf changes.

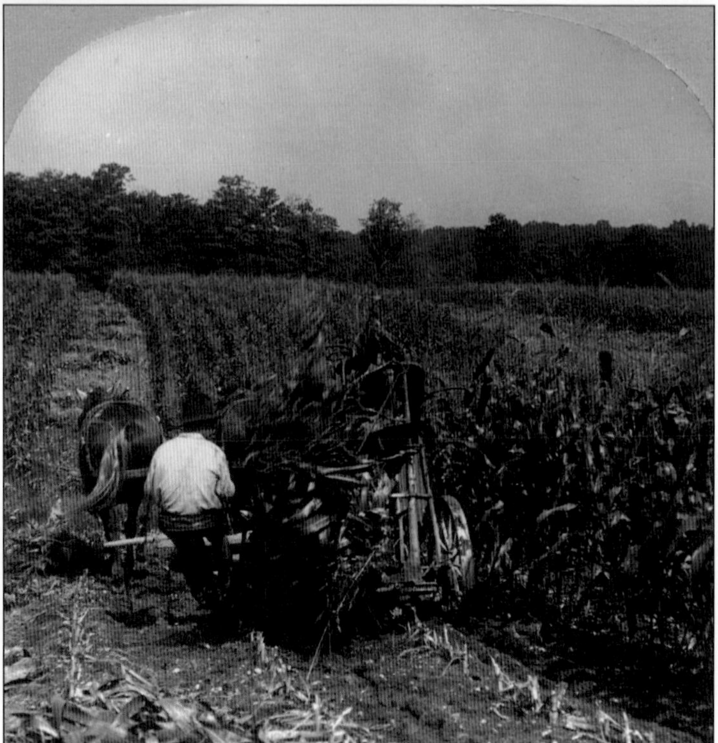

Man on Horse Harvesting. Pictured here is one of the many men in Andrew Squire's employ. Squire took great pride in his crops, his animals, and his farm. Whatever part of this harvest was not destined for the Squires' dining room most likely went to the Union Club in downtown Cleveland. The club was a big purchaser of Valleevue Farm's food.

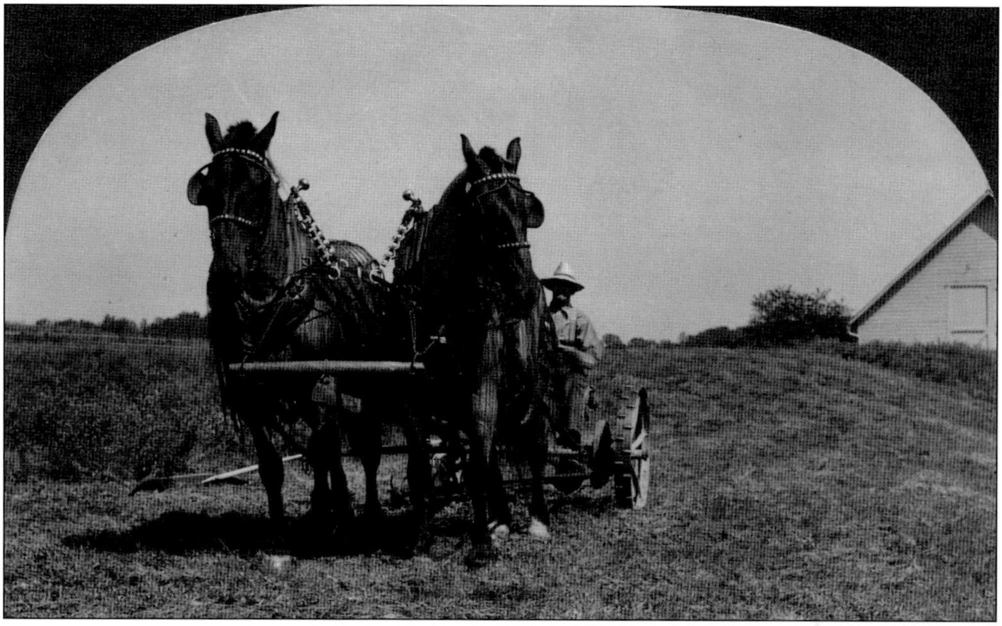

Man Driving Horses in Field. Some of Andrew Squire's handsome workhorses are seen being driven by one of his hired hands. Not having to rely solely on the income from the sale of his crops, Squire made it his passion to purchase some of the finest breeds of animal available to work his land. Not many farmers, then or now, would be envious of his high overhead.

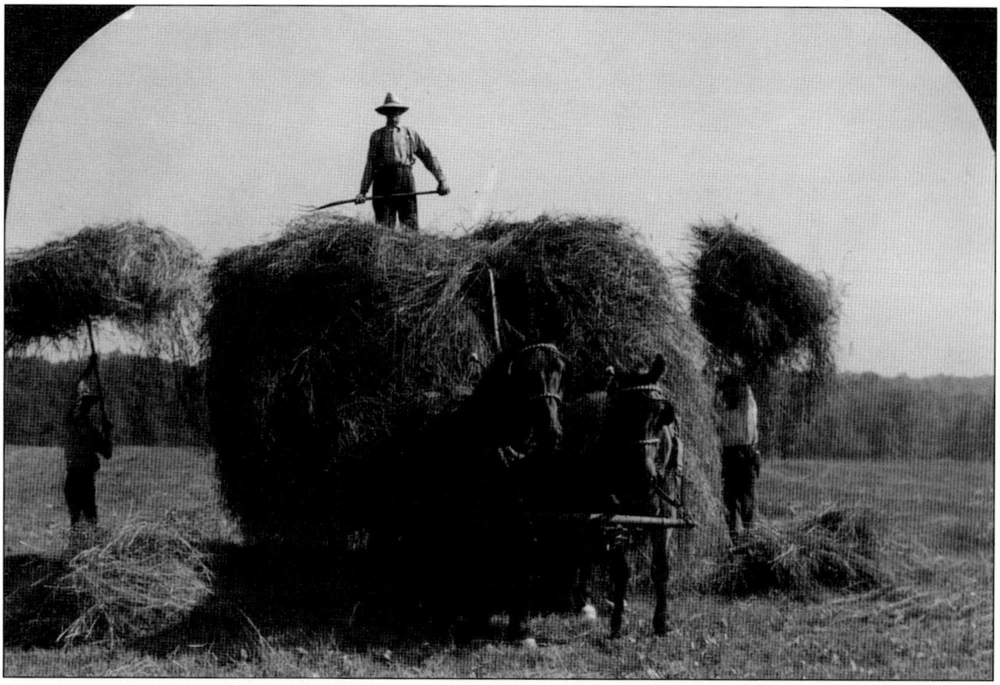

Men Harvesting Hay. Andrew Squire's farmhands are pictured engaging in the perennial task of harvesting the hay. Keeping the animals' bellies full was no less an enormous task than feeding all of the employees. The scores of animals residing at Valleevue Farm were likely very well fed.

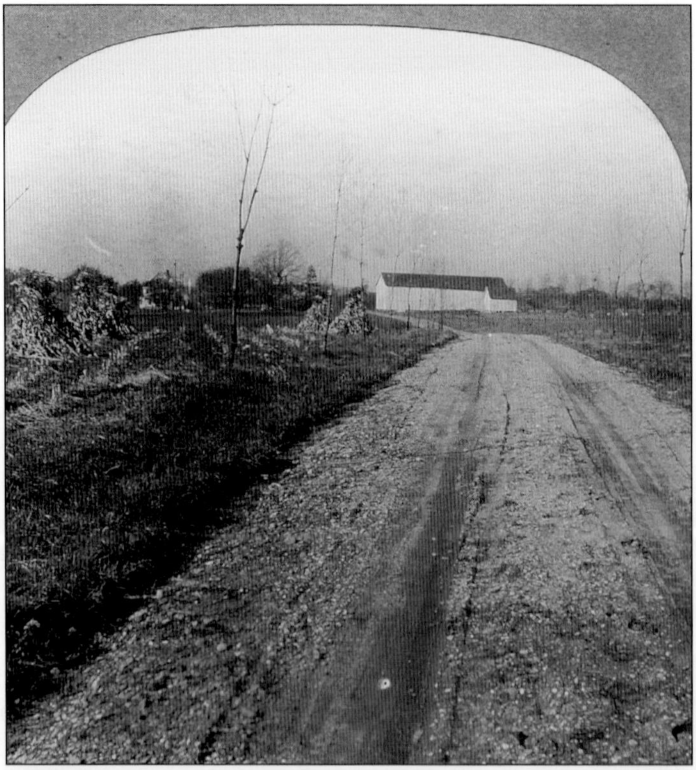

Bundled Corn Stalks. In these 1925 photographs, cornstalks are seen in the fields in their postharvest bundles (left) off of the road connecting the main farm to the farmhouse on S.O.M. Center Road and (below) behind a split-rail fence in the upper field to the west of what is now the main barn complex.

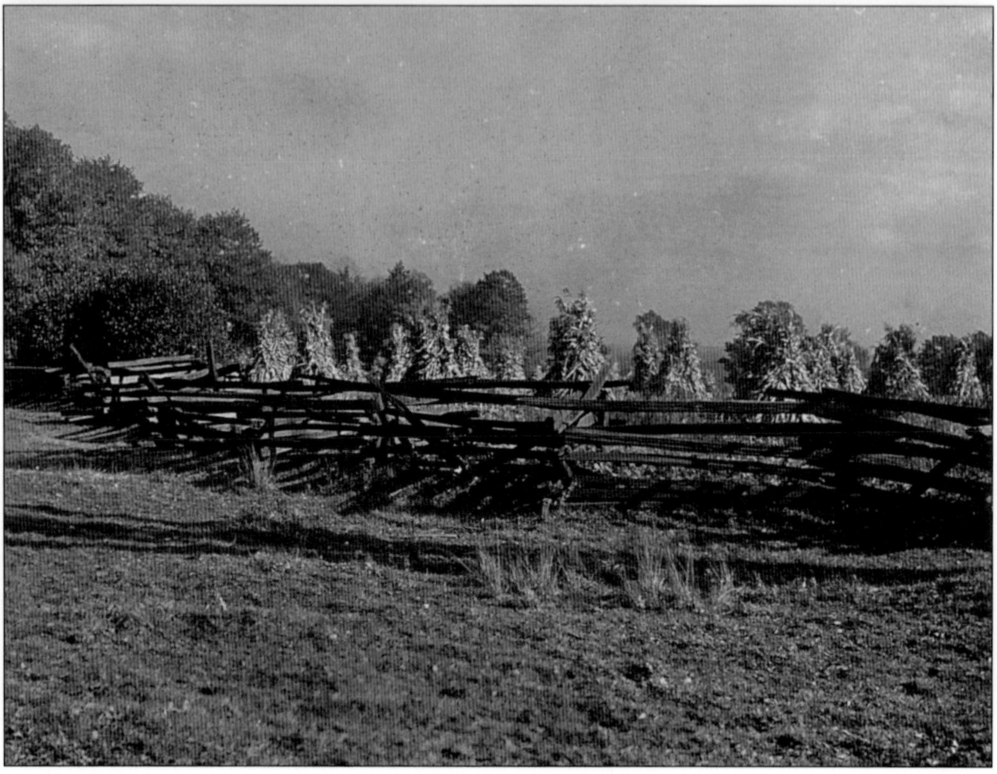

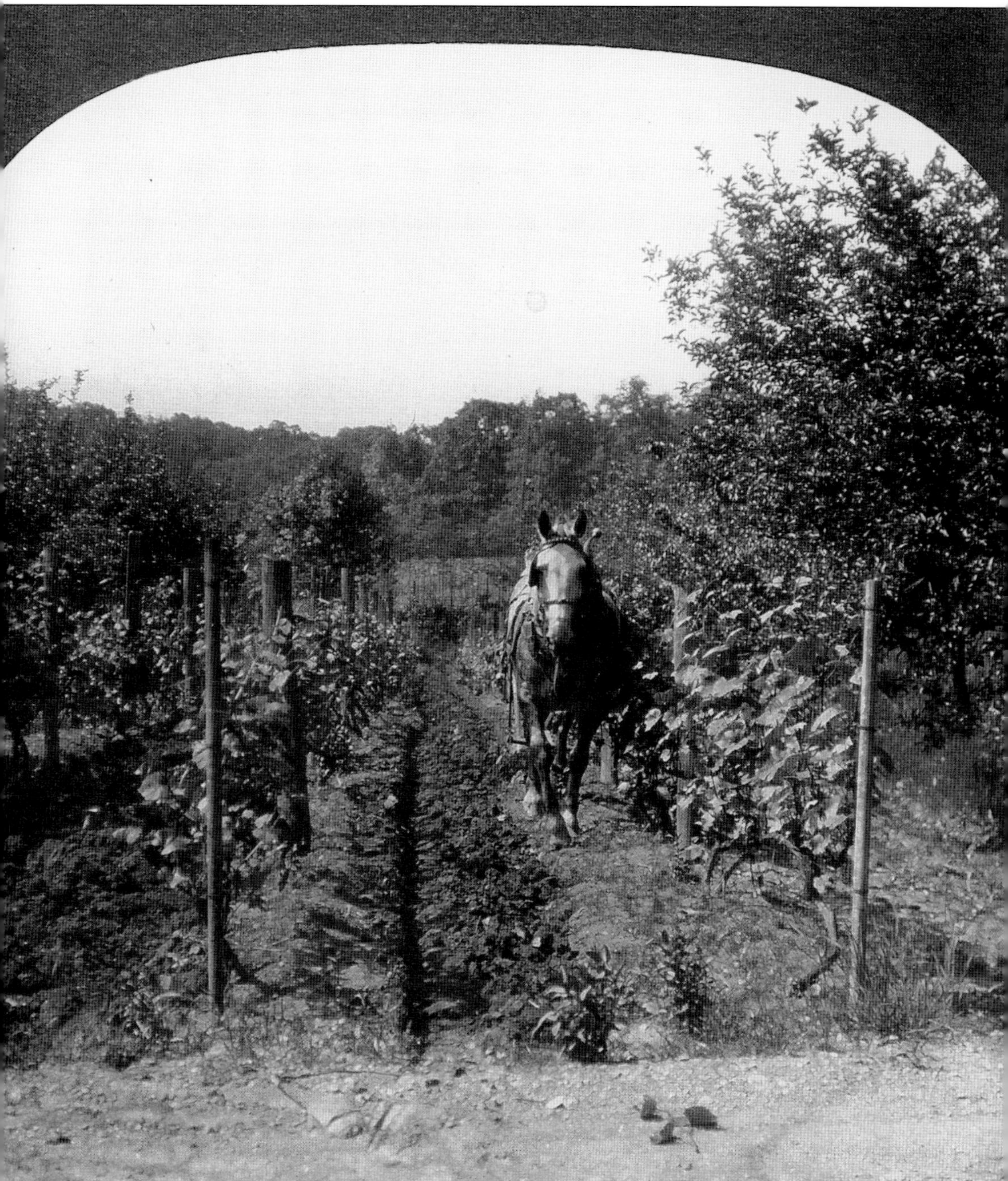

Plucking Grapes. A farmhand collects grape clusters from a horse-pulled cart. The grapevines at Valleevue Farm were planted nearly 100 years ago and are still producing Concord grapes to this day. Grapes from these old vines and their progeny are still harvested every year, although horses are no longer used. Over the years, some have been transplanted to other areas of the farm, and some have been lost to disease, insects, or old age. Some of the grapes from this original site can be seen next to the driveway that leads to the present-day Sheep Barn.

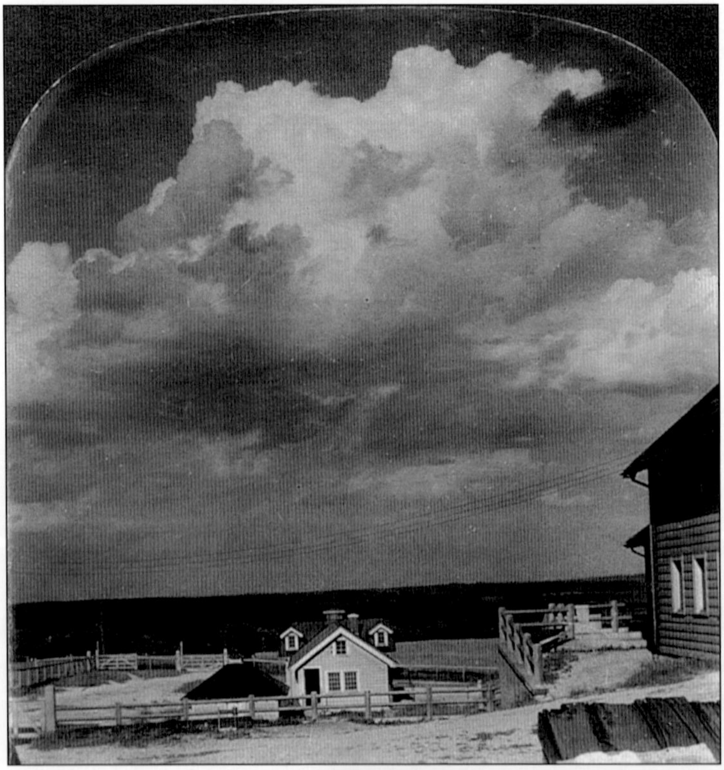

The Pig Barn on a Cloudy Day. This is the only known photograph of the pig barn that does not depict it in use by the animals of the farm during the time of Andrew Squire. The image was presumably taken shortly after the barn's completion. The smaller building in front and to the left of the pig barn houses a cistern. The portion of the building that is seen to the right is the main barn.

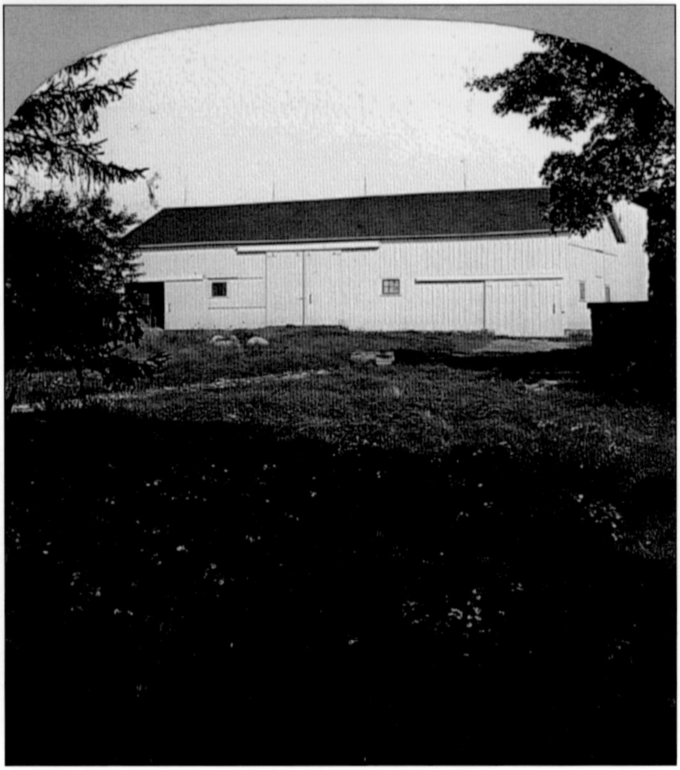

The Horse Barn. The horse barn, as it was originally used, would later serve as a sheep barn during the time that the Flora Stone Mather College used the farm for its agricultural activities. It was located between the main barn and the greenhouse.

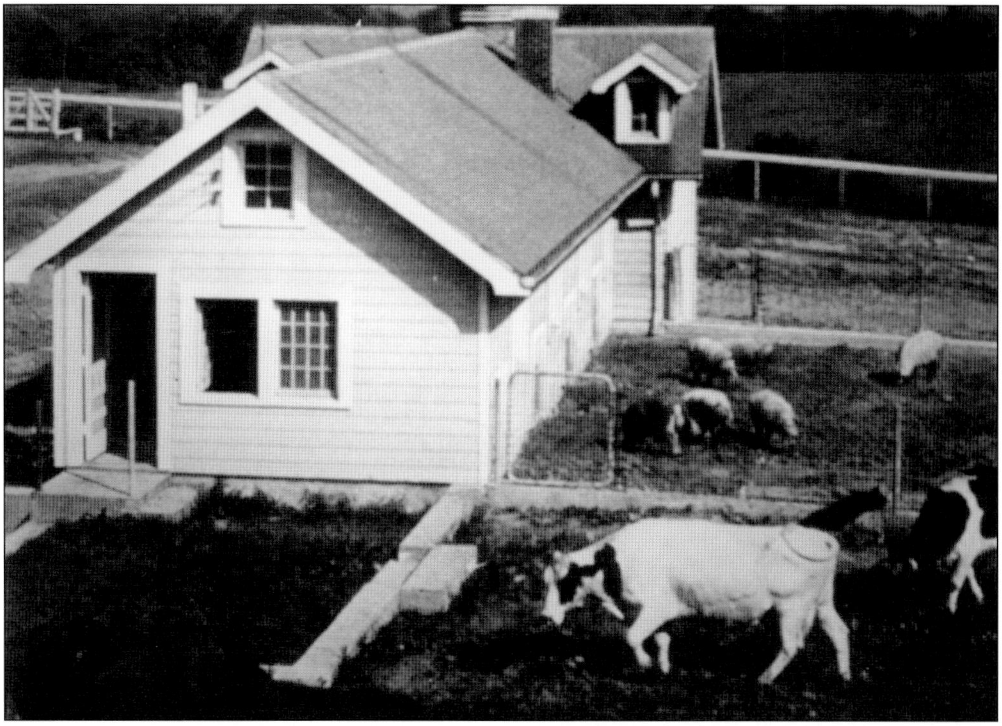

PIGS AND COWS BY PIG BARN. Here, in its original manifestation, is the pig barn, later to be dubbed "The Pink Pig" by Flora Stone Mather College students. Like his neighbor to the east, Jeptha Homer Wade II, Andrew Squire had some of his barns and buildings designed by a prominent architectural firm. The building pictured in the right-hand side of the picture below is the cistern.

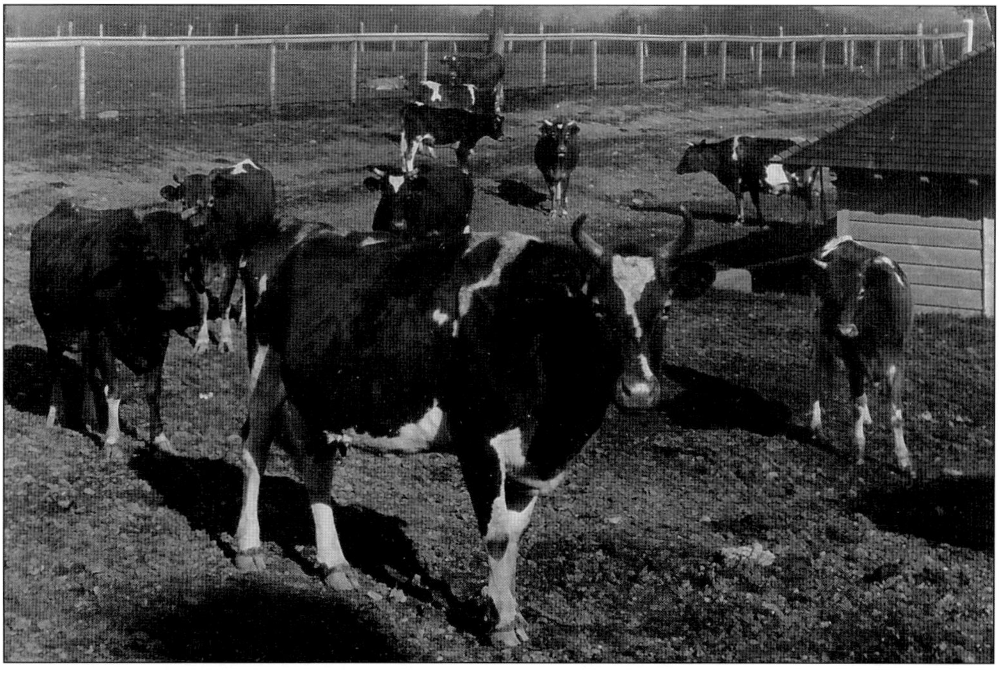

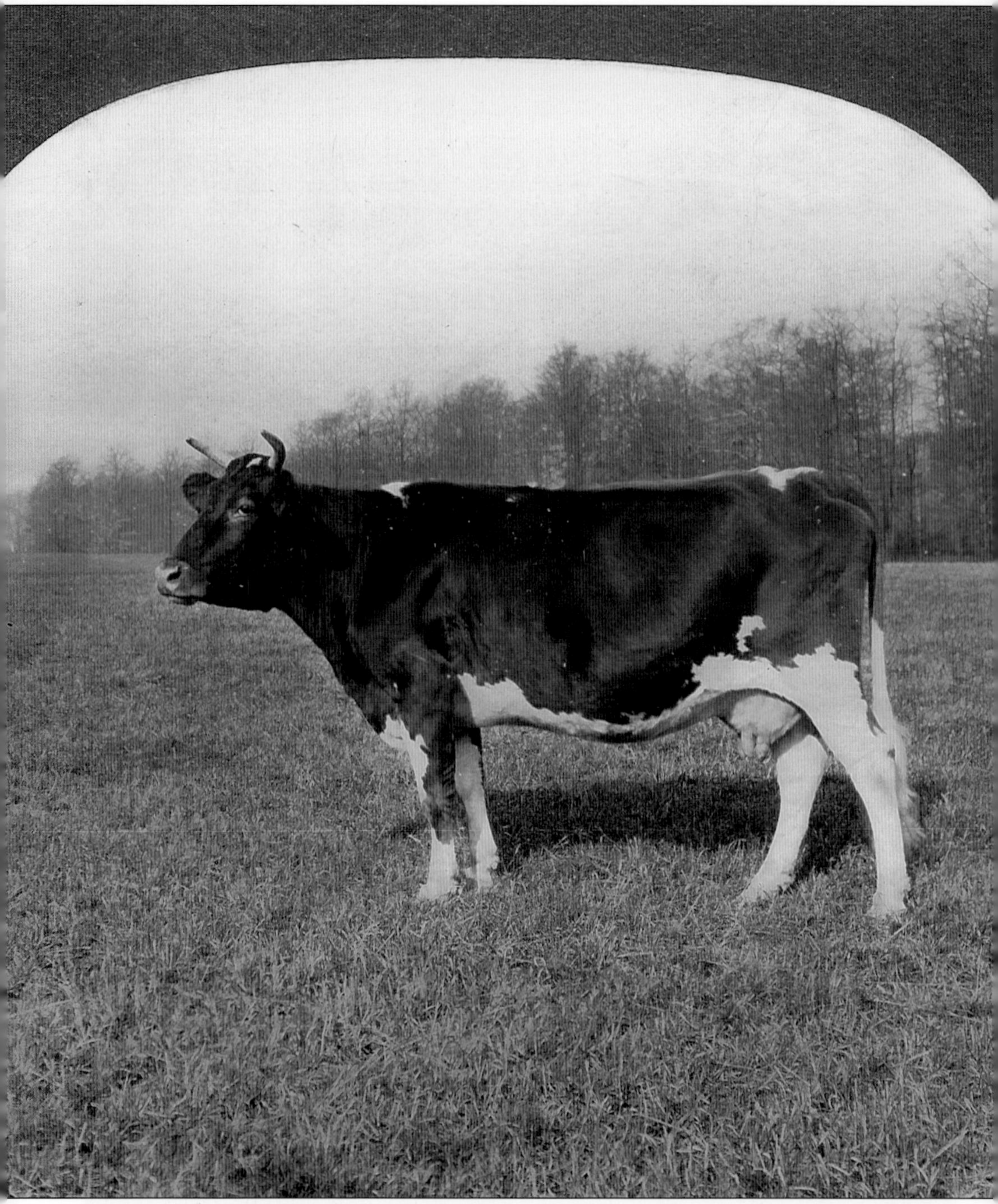

Cow. In this stereoscopic image from the early 20th century is one of Andrew Squire's cows. Other images that exist of his cows show their tails braided. Along with many of the other estate farmers in Hunting Valley who did not need to derive their income from the practice of farming, Squire had the luxury of engaging in highly selective breeding on his farm. Eugenics was a common practice with livestock. The appearance of his animals was of high import to him.

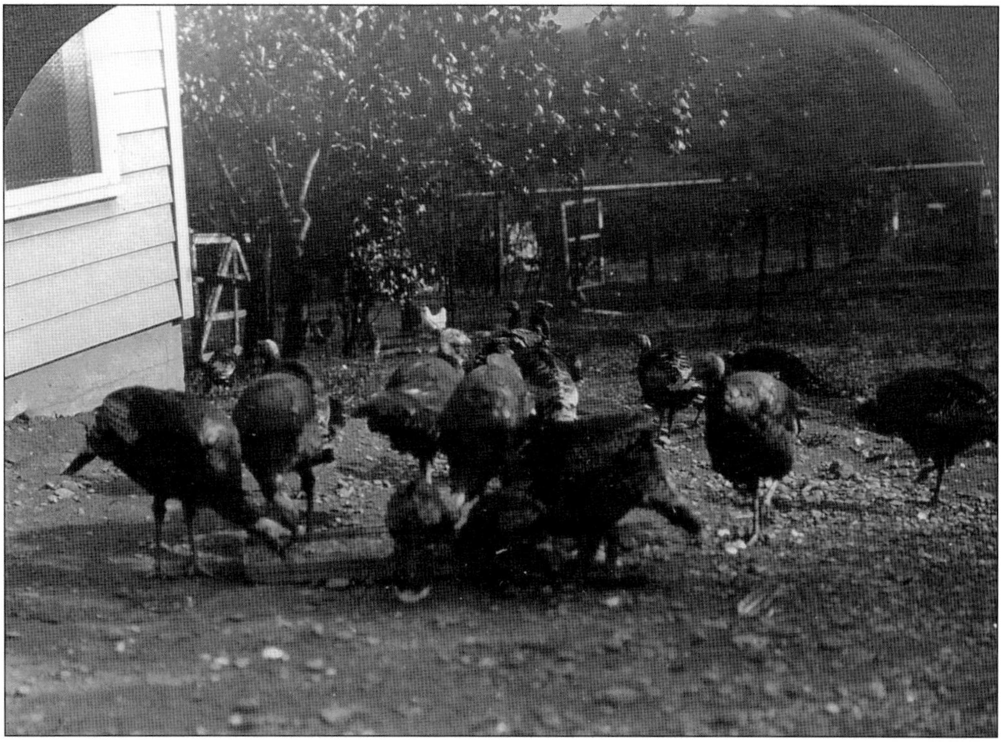

Turkeys and Ducks. In addition to large numbers of cows, pigs, and horses, Andrew Squire stocked his farm with a wide variety of fowl, including these turkeys and ducks that could be found waddling about around the barnyards.

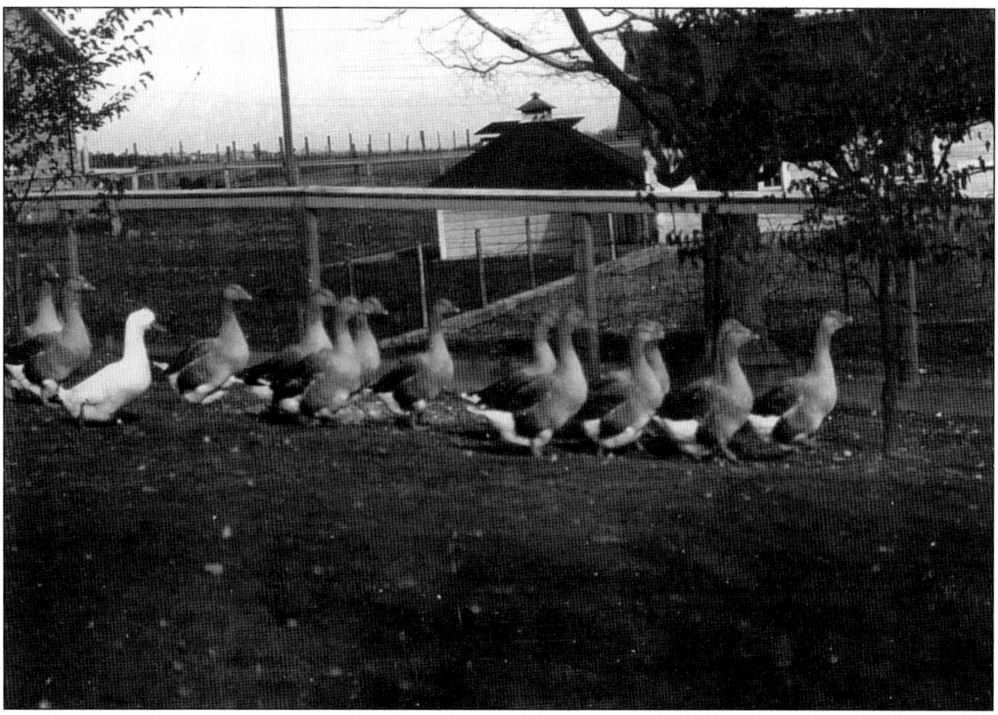

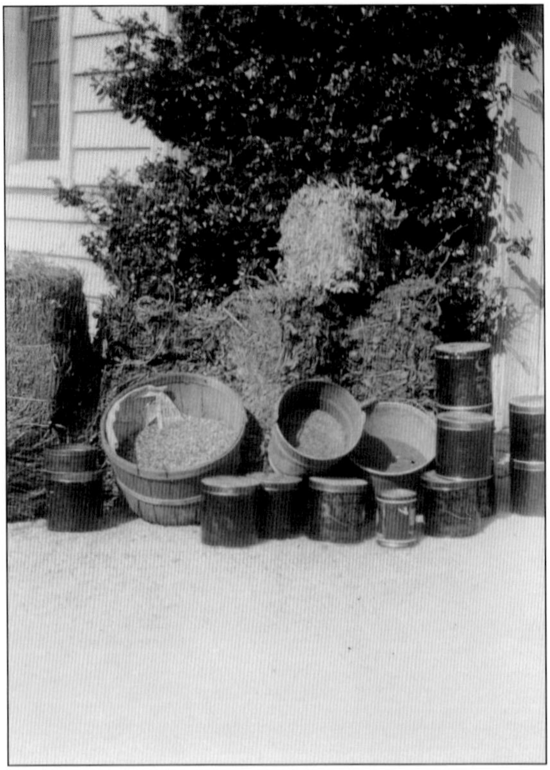

PRODUCTS FROM THE PHARMACEUTICAL GARDEN. Before passing the bar exam, Andrew Squire's first professional aspiration was to be a doctor, like his father. He even attended medical school. His vast estate allowed him to dabble in healing the sick by installing a pharmaceutical garden from which compounds and tinctures were made to supply local institutions, including the university hospital and its heart patients.

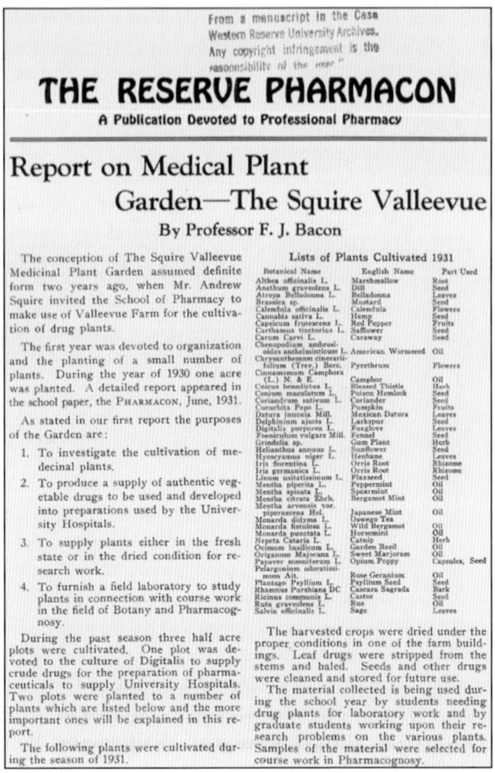

THE RESERVE PHARMACON
A Publication Devoted to Professional Pharmacy

Report on Medical Plant Garden—The Squire Valleevue

By Professor F. J. Bacon

The conception of The Squire Valleevue Medicinal Plant Garden assumed definite form two years ago, when Mr. Andrew Squire invited the School of Pharmacy to make use of Valleevue Farm for the cultivation of drug plants.

The first year was devoted to organization and the planting of a small number of plants. During the year of 1930 one acre was planted. A detailed report appeared in the school paper, the PHARMACON, June, 1931.

As stated in our first report the purposes of the Garden are:

1. To investigate the cultivation of medecinal plants.
2. To produce a supply of authentic vegetable drugs to be used and developed into preparations used by the University Hospitals.
3. To supply plants either in the fresh state or in the dried condition for research work.
4. To furnish a field laboratory to study plants in connection with course work in the field of Botany and Pharmacognosy.

During the past season three half acre plots were cultivated. One plot was devoted to the culture of Digitalis to supply crude drugs for the preparation of pharmaceuticals to supply University Hospitals. Two plots were planted to a number of plants which are listed below and the more important ones will be explained in this report.

The following plants were cultivated during the season of 1931.

Lists of Plants Cultivated 1931

Botanical Name	English Name	Part Used
Althea officinalis L.	Marshmallow	Root
Anethum graveolens L.	Dill	Seed
Atropa Belladonna L.	Belladonna	Leaves
Brassica sp.	Mustard	Seed
Calendula officinalis L.	Calendula	Flowers
Cannabis sativa L.	Hemp	Seed
Capsicum frutescens L.	Red Pepper	Fruits
Carthamus tinctorius L.	Safflower	Seed
Carum Carvi L.	Caraway	Seed
Chenopodium ambrosioides anthelminticum L.	American Wormseed	Oil
Chrysanthemum cinerariifolium (Trev.) Borc.	Pyrethrum	Flowers
Cinnamomum Camphora (L.) N. & E.	Camphor	Oil
Cnicus benedictus L.	Blessed Thistle	Herb
Conium maculatum L.	Poison Hemlock	Seed
Coriandrum sativum L.	Coriander	Seed
Cucurbita Pepo L.	Pumpkin	Fruits
Datura innoxia Mill.	Mexican Datura	Leaves
Delphinium ajacis L.	Larkspur	Seed
Digitalis purpurea L.	Foxglove	Leaves
Foeniculum vulgare Mill.	Fennel	Seed
Grindelia sp.	Gum Plant	Herb
Helianthus annuus L.	Sunflower	Seed
Hyoscyamus niger L.	Henbane	Leaves
Iris florentina L.	Orris Root	Rhizome
Iris germanica L.	Orris Root	Rhizome
Linum usitatissimum L.	Flaxseed	Seed
Mentha piperita L.	Peppermint	Oil
Mentha spicata L.	Spearmint	Oil
Mentha citrata Ehrh.	Bergamot Mint	Oil
Mentha arvensis var. piperascens Hol.	Japanese Mint	Oil
Monarda didyma L.	Oswego Tea	
Monarda fistulosa L.	Wild Bergamot	Oil
Monarda punctata L.	Horsemint	Oil
Nepeta Cataria L.	Catnip	Herb
Ocimum basilicum L.	Garden Basil	Oil
Origanum Majorana L.	Sweet Marjoram	Oil
Papaver somniferum L.	Opium Poppy	Capsules, Seed
Pelargonium odoratissimum Ait.	Rose Geranium	Oil
Plantago Psyllium L.	Psyllium Seed	Seed
Rhamnus Purshiana DC	Cascara Sagrada	Bark
Ricinus communis L.	Castor	Seed
Ruta graveolens L.	Rue	Oil
Salvia officinalis L.	Sage	Leaves

The harvested crops were dried under the proper conditions in one of the farm buildings. Leaf drugs were stripped from the stems and baled. Seeds and other drugs were cleaned and stored for future use.

The material collected is being used during the school year by students needing drug plants for laboratory work and by graduate students working upon their research problems on the various plants. Samples of the material were selected for course work in Pharmacognosy.

THE FIRST REPORT ON THE PHARMACEUTICAL GARDEN. Pictured is a page from a 1934 report by Andrew Squire's friend and Western Reserve University faculty member Dr. Franklin J. Bacon. This report highlighted the production in Squire's medicinal gardens and the various uses of the products grown within. Dr. Bacon became Valleevue Farm's first director upon the transfer of the property to the university. He and his family were the Manor House's first residents after the passing of Eleanor Squire.

PORTION OF ESTIMATE BY WARREN MANNING. At the insistence of the neighboring Wade family, noted American landscape architect Warren Manning visited the estate numerous times to develop a landscape plan. Working with the architects Walker and Weeks, Manning proposed a landscape estimate in 1913 that would total $30,912.58.

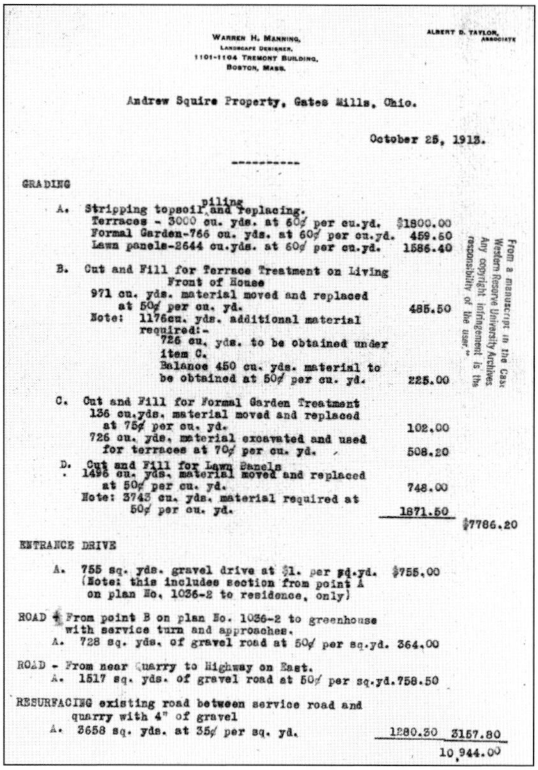

WARREN H. MANNING,
LANDSCAPE DESIGNER,
1101-1104 TREMONT BUILDING,
BOSTON, MASS.

ALBERT D. TAYLOR,
ASSOCIATE

Andrew Squire Property, Gates Mills, Ohio.

October 25, 1913.

GRADING

A. Stripping topsoil, piling and replacing.
Terraces - 3000 cu. yds. at 60¢ per cu.yd. $1800.00
Formal Garden-766 cu. yds. at 60¢ per cu.yd. 459.60
Lawn panels-2644 cu.yds. at 60¢ per cu.yd. 1586.40

B. Cut and Fill for Terrace Treatment on Living Front of House
971 cu. yds. material moved and replaced at 50¢ per cu. yd. 485.50
Note: 1176cu. yds. additional material required:-
726 cu. yds. to be obtained under item C.
Balance 450 cu. yds. material to be obtained at 50¢ per cu. yd. 225.00

C. Cut and Fill for Formal Garden Treatment
136 cu.yds. material moved and replaced at 75¢ per cu. yd. 102.00
726 cu. yds. material excavated and used for terraces at 70¢ per cu. yd. 508.20

D. Cut and Fill for Lawn Panels
1496 cu. yds. material moved and replaced at 50¢ per cu. yd. 748.00
Note: 3743 cu. yds. material required at 50¢ per cu. yd. 1871.50

$7786.20

ENTRANCE DRIVE

A. 755 sq. yds. gravel drive at $1. per sq.yd. $755.00
(Note: this includes section from point A on plan No. 1036-2 to residence, only)

ROAD - From point B on plan No. 1036-2 to greenhouse with service turn and approaches.
A. 728 sq. yds. of gravel road at 50¢ per sq.yd. 364.00

ROAD - From near Quarry to Highway on East.
A. 1517 sq. yds. of gravel road at 50¢ per sq.yd. 758.50

RESURFACING existing road between service road and quarry with 4" of gravel
A. 3658 sq. yds. at 35¢ per sq. yd. 1280.30 3157.80

10,944.00

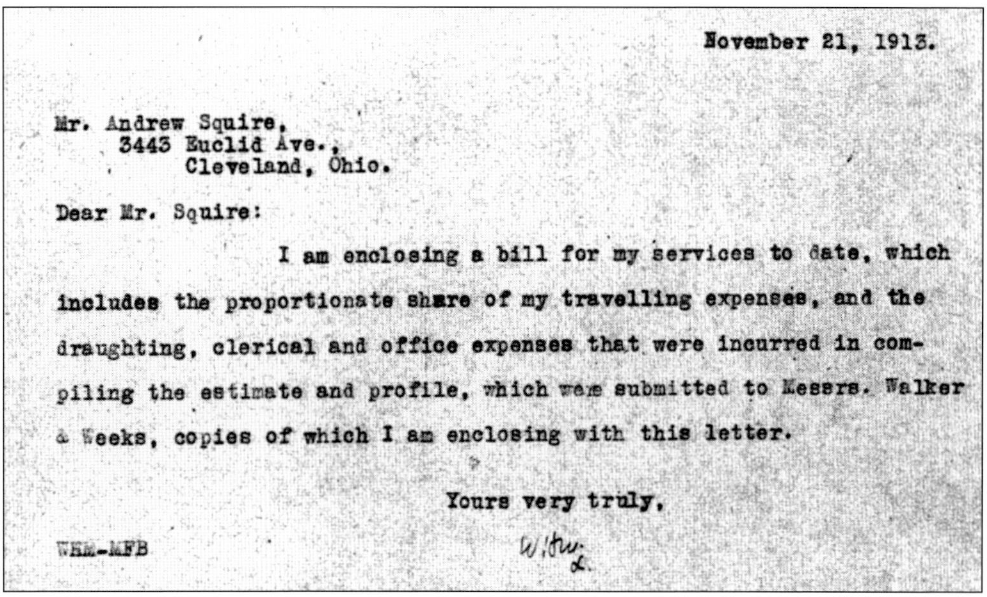

November 21, 1913.

Mr. Andrew Squire,
3443 Euclid Ave.,
Cleveland, Ohio.

Dear Mr. Squire:

I am enclosing a bill for my services to date, which includes the proportionate share of my travelling expenses, and the draughting, clerical and office expenses that were incurred in compiling the estimate and profile, which were submitted to Messrs. Walker & Weeks, copies of which I am enclosing with this letter.

Yours very truly,

WHM-MFB

LETTER FROM MANNING TO SQUIRES. A bill for Warren Manning's services to date was presented to the Squires for all of his survey and estimating work. A portion is seen here. Eleanor Squire had decided that Manning was not a suitable designer because he was located in Boston and, therefore, would be unable to personally supervise the work. Manning also had artistic differences with Walker and Weeks, much preferring the architectural style of Hubbell and Benes.

3 4 4 3 Euclid Avenue

Cleveland, Ohio, June 2, 1914.

Mr. Warren H. Manning,

1101-1104 Tremont Building,

Boston, Mass.

RECEIVED June 6, 1914 WARREN H. MANNING

My dear Mr. Manning:-

I asked Mr. Squire to allow me to reply to your letter and its enclosure, but as the sort of thing I have to say is difficult to write, I have put it off from month to month; but as you very likely know, I am now confined to my room with another broken bone, so that I have plenty of time to go into details.

Mr. Squire and I have made up our minds that, with all the garden training of the past few years, no one can lay out our farm or advise me on the planning who would not be there with me much oftener than two or three times a year. Therefore, please permit me to enclose check for $100.00 to cover your charge for your visit which you so kindly made to us for the farm last fall.

In regard to the charges which were made by your office and which have been increased a little with each presentation of the bill, we do not feel that you would wish us to pay these charges if you really understood the facts in the case. Your office presented its estimates and bids on work totally and entirely in the hands of another firm. The only possible excuse for the presentation of such bills we can see is, that they must have gone through the routine of your office on the arrival of a plan sent from Walker & Weeks of Cleveland before your Mr. Taylor, who went west, had explained to your office why he had asked to have such plans sent. I have asked Walker & Weeks if they requested your office to make

Letter from Eleanor Squire to Warren Manning. In a portion of Eleanor Squire's response to Warren Manning dated June 2, 1914, she writes, "we do not feel that you would wish us to pay these charges if you really understood the facts in the case." Her contention was that plans were made that were not requested and that those that were requested were merely plans and not estimates. Therefore, she should not have been charged for the estimate.

Written personally by W.H.M.

December 7,1915.

Dear Mr. Squire,

I find that there is $76.20 due on my account for the visit and report I made on your estate near Gates Mills after the visit there with you and Mrs. Squire. I presume the matter has been overlooked. Could I ask you to look into this and give it your attention? I shall appreciate this.

Yours truly,
W.H.M.

Letter from Manning to Squire. The contest over Manning's bill would go on for some time. In a brief letter from Manning dated December 7, 1915, he assumes that "the matter [of his bill not being paid] has been overlooked" and asks Squire "to look into this and give it your attention." There is no evidence that this balance was ever paid, and this is the last known correspondence on the matter.

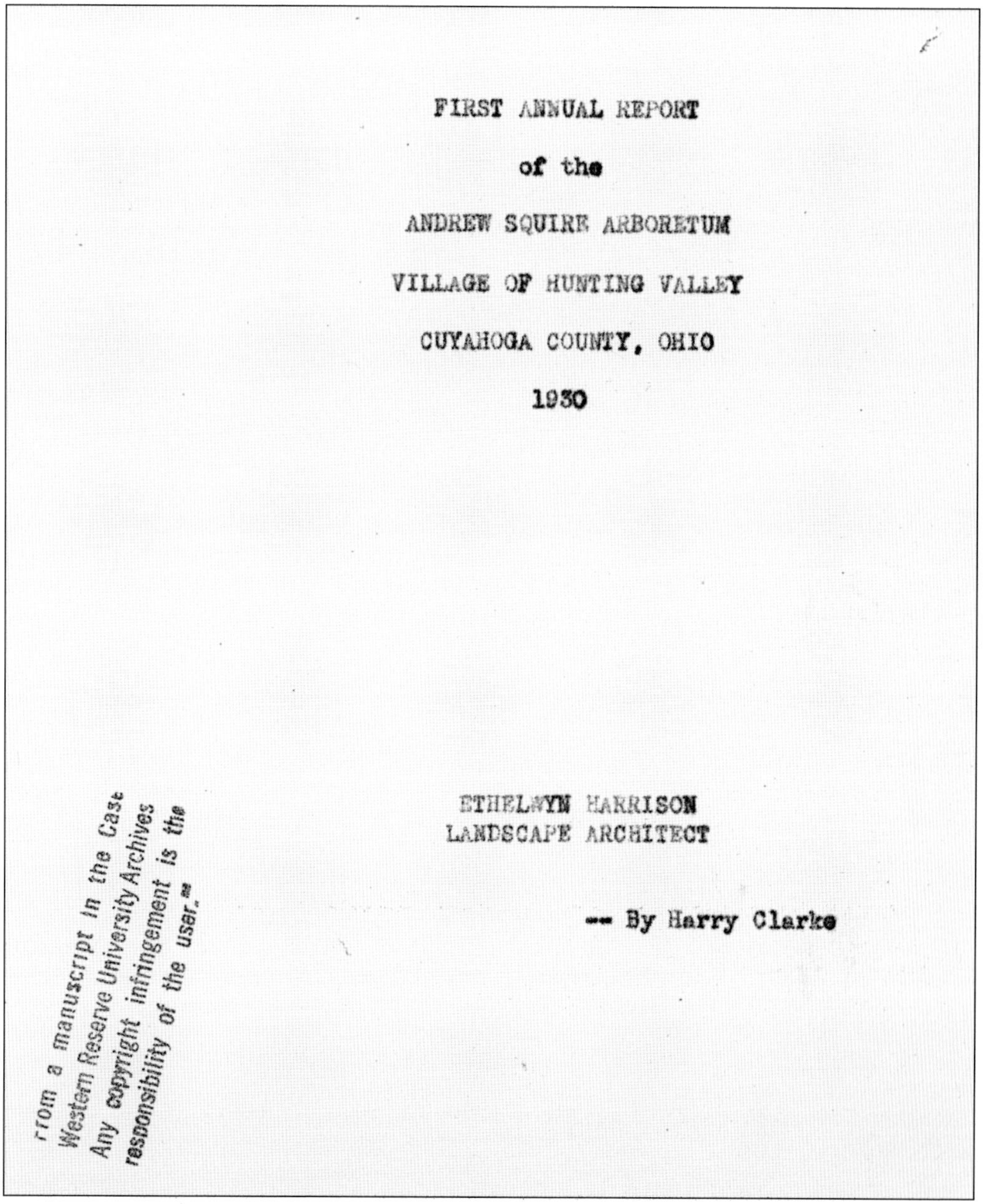

FIRST ANNUAL REPORT

of the

ANDREW SQUIRE ARBORETUM

VILLAGE OF HUNTING VALLEY

CUYAHOGA COUNTY, OHIO

1930

ETHELWYN HARRISON
LANDSCAPE ARCHITECT

— By Harry Clarke

Partial Inventory of Squire Arboretum. After it was apparent that the Squires would not be working with Warren Manning, Andrew Squire hired Cleveland area landscape architect Violet Ethelwyn Harrison (who also served as the farm's superintendent until Eleanor Squire's death in 1937) to develop a 900-species arboretum in the southwest portion of his estate, with representation of every known native variety of florae that could be obtained. Harrison would prepare annual reports on the Squire Arboretum, detailing what was installed that year and updating the arboretum's inventory. These lengthy reports included such information as each plant's habit and bloom time. They were generated throughout the 1930s for Andrew Squire and his estate. The reports ceased upon transfer of the property to the university because Harrison did not remain employed there. She continued to pursue her own landscape architecture business.

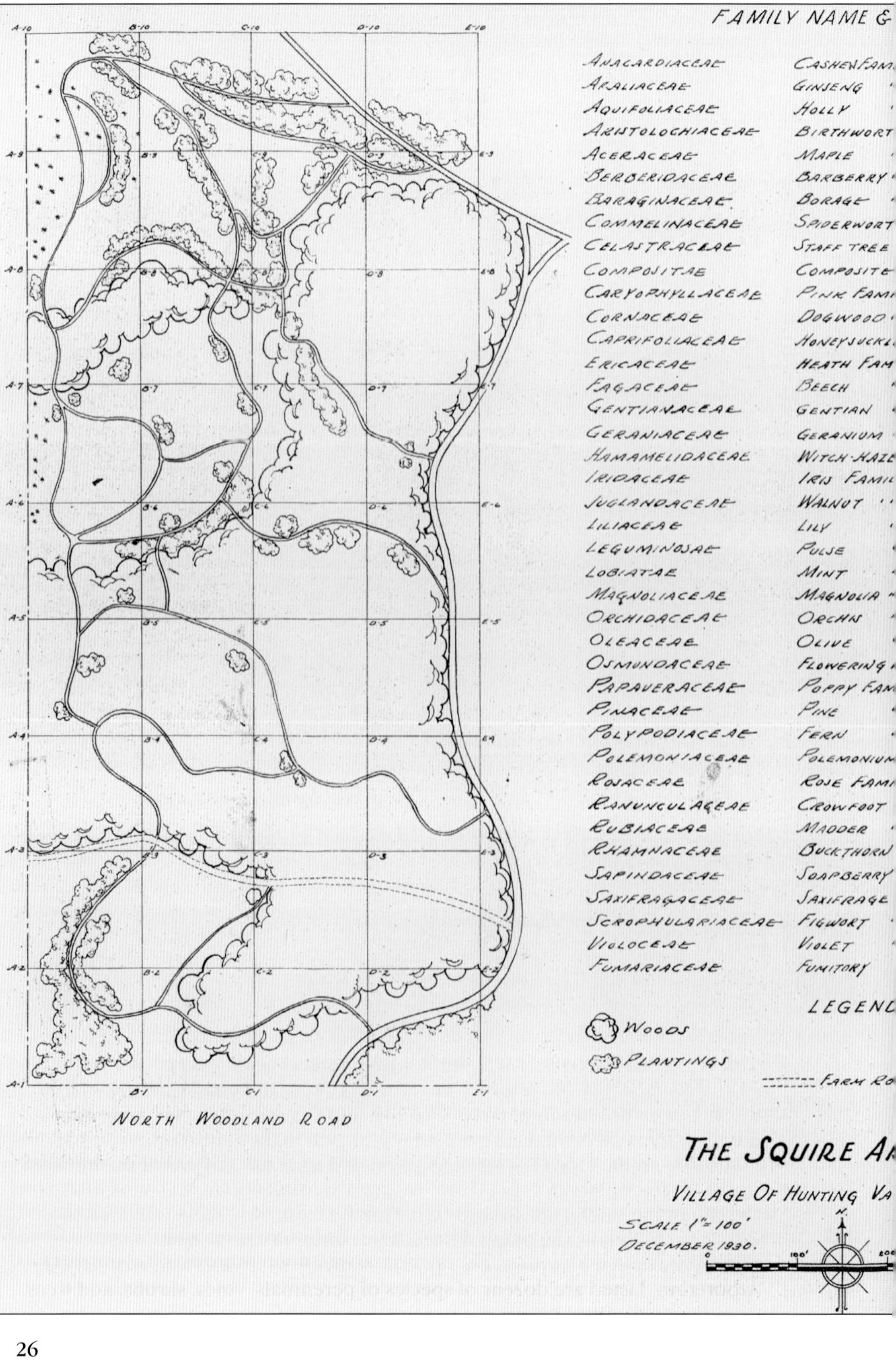
FAMILY NAME &
ANACARDIACEAE CASHEW FAM
ARALIACEAE GINSENG
AQUIFOLIACEAE HOLLY
ARISTOLOCHIACEAE BIRTHWORT
ACERACEAE MAPLE
BERBERIDACEAE BARBERRY
BARAGINACEAE BORAGE
COMMELINACEAE SPIDERWORT
CELASTRACEAE STAFF TREE
COMPOSITAE COMPOSITE
CARYOPHYLLACEAE PINK FAM
CORNACEAE DOGWOOD
CAPRIFOLIACEAE HONEYSUCKL
ERICACEAE HEATH FAM
FAGACEAE BEECH
GENTIANACEAE GENTIAN
GERANIACEAE GERANIUM
HAMAMELIDACEAE WITCH-HAZE
IRIDACEAE IRIS FAMIL
JUGLANDACEAE WALNUT
LILIACEAE LILY
LEGUMINOSAE PULSE
LOBIATAE MINT
MAGNOLIACEAE MAGNOLIA
ORCHIDACEAE ORCHID
OLEACEAE OLIVE
OSMUNDACEAE FLOWERING
PAPAVERACEAE POPPY FAM
PINACEAE PINE
POLYPODIACEAE FERN
POLEMONIACEAE POLEMONIUM
ROSACEAE ROSE FAMI
RANUNCULACEAE CROWFOOT
RUBIACEAE MADDER
RHAMNACEAE BUCKTHORN
SAPINDACEAE SOAPBERRY
SAXIFRAGACEAE SAXIFRAGE
SCROPHULARIACEAE FIGWORT
VIOLOCEAE VIOLET
FUMARIACEAE FUMITORY
LEGEND
WOODS
PLANTINGS
FARM RO
NORTH WOODLAND ROAD
THE SQUIRE AR
VILLAGE OF HUNTING VA
SCALE 1" = 100'
DECEMBER 1930.
N.

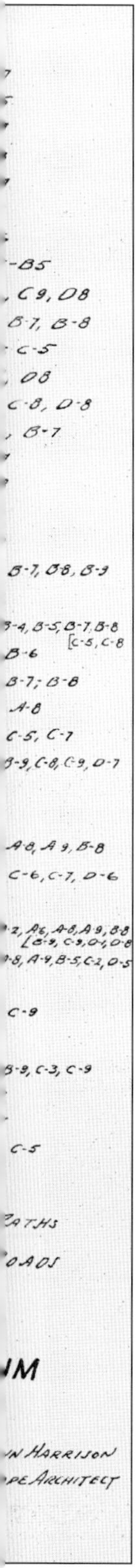

The Squire Arboretum. The Squires would invest a great deal of time and money in their beloved arboretum and would entrust Ethelwyn Harrison (she did not often use her first name, Violet) to tend to its every need. After her time at Valleevue Farm, Harrison (who is also thought to be the first female landscape architect in Ohio) would go on to design the gardens of the Blossom Estate in nearby Beachwood, Ohio. Legacy Village Shopping Center now sits on the site of the former Blossom Estate. This 1930 plan in Ethelwyn Harrison's handwriting details the original florae by taxonomic family name and common name and outlines each plant's respective location within the Squire Arboretum. Listed are dozens of species of perennials, vines, shrubs, and trees, most of which are now long gone.

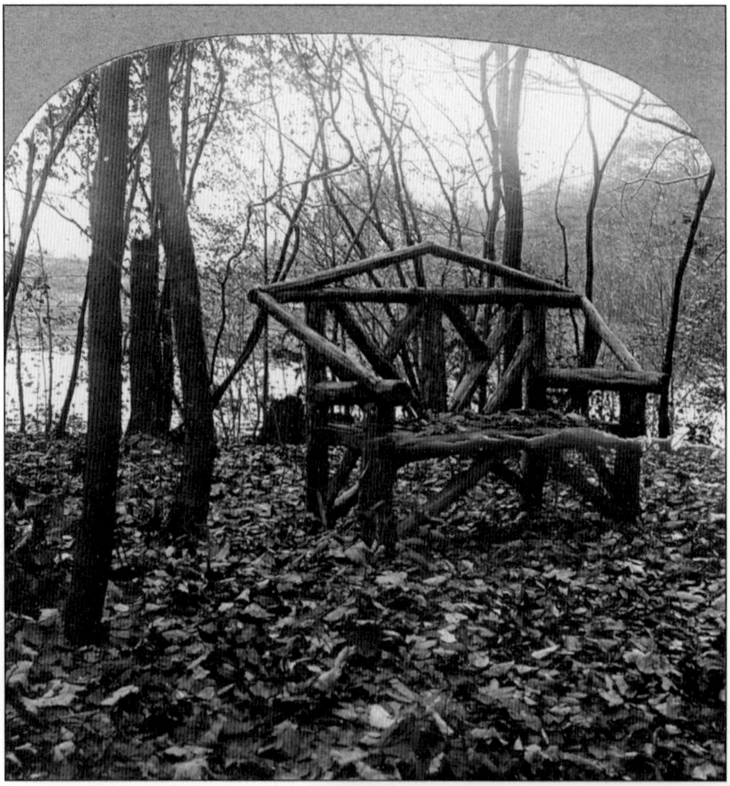

A Place to Sit. Hidden among the trails and coves of the arboretum were several spots to stop and reflect upon the beauty all around. Wooden benches like the one seen here were strewn about the arboretum for guests and Andrew Squire's little black dog alike.

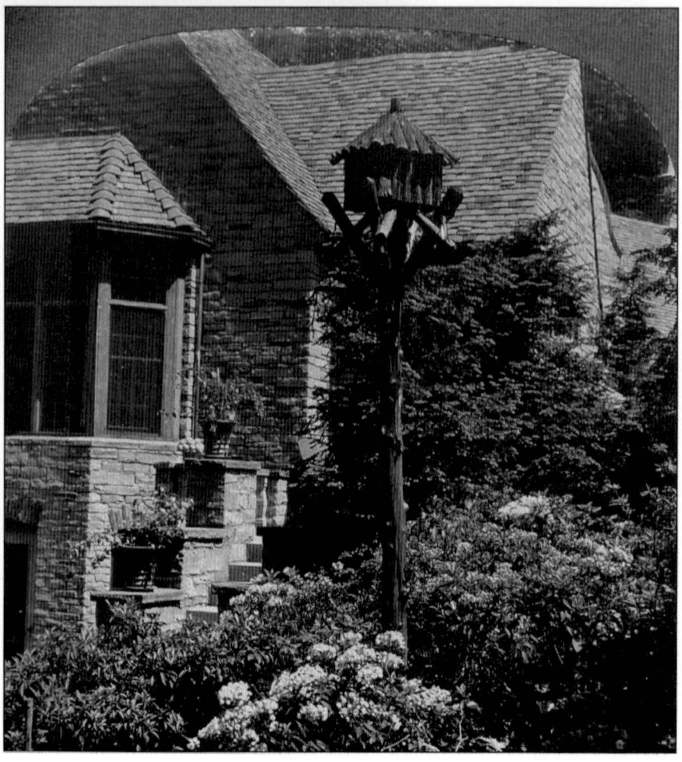

Birdhouse. Catering to the fauna that existed at Valleevue Farm was as important as tending to its flora. This unique birdhouse, situated just outside of the Manor House, would have offered residence to a few of the thousands of birds that called Squire Valleevue Farm home. Birdhouses and birdbaths were common accents around the Manor House in the Squire's day. Valleevue Farm continues to host over 200 distinct species of native and migratory birds within its boundaries.

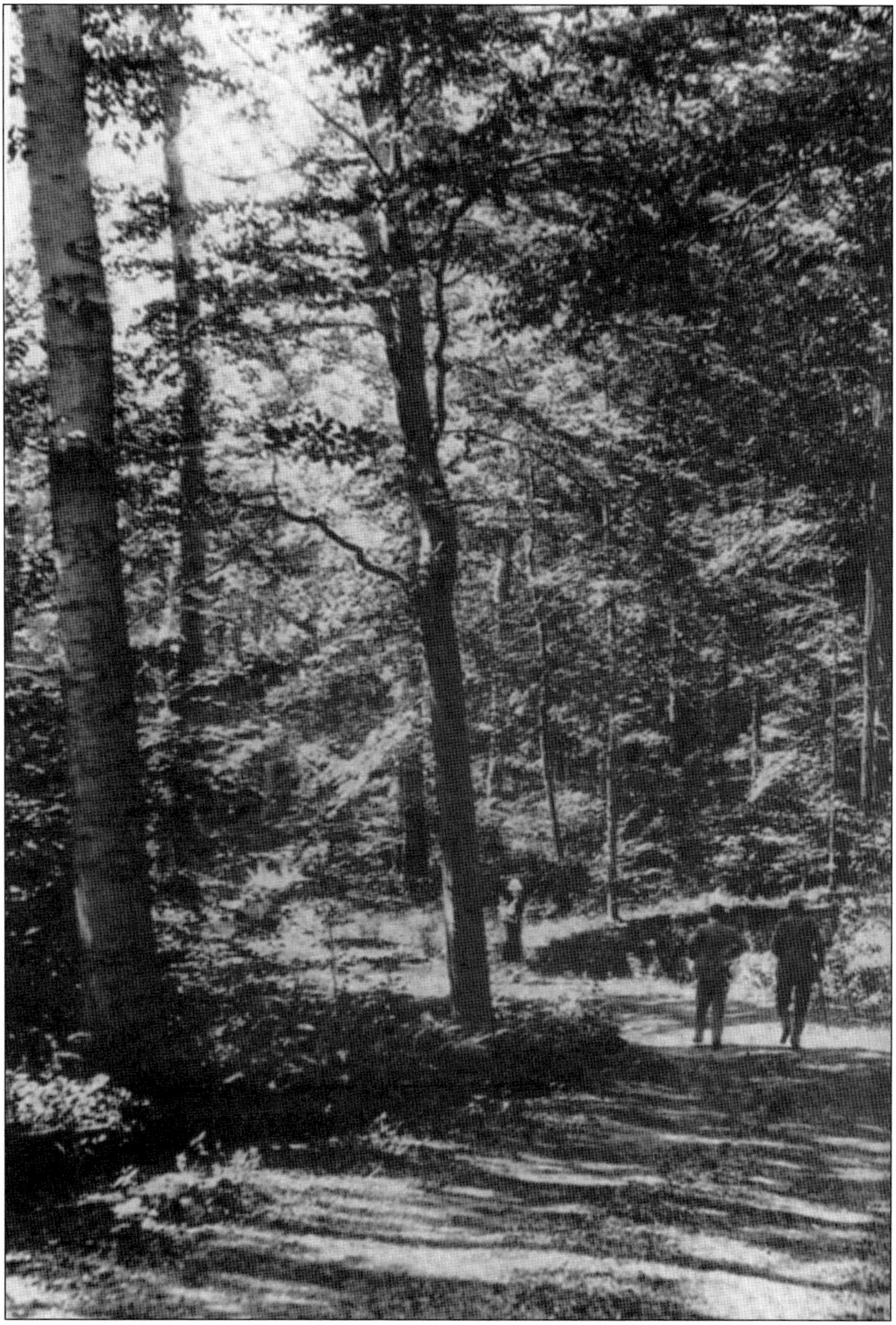

ANDREW SQUIRE HIKING. Andrew Squire is pictured "taking the air" with an unidentified friend on one of the many tanbark (mulch) paths that wound into, out of, and around the Squire Arboretum. Squire was quite proud of the arboretum and often would take his guests on hikes through it. The Squires spent many years perfecting the site before building their home there. It is said that one reason for the delay in constructing their home was an inability to decide which vantage point would be most advantageous for the location of the house.

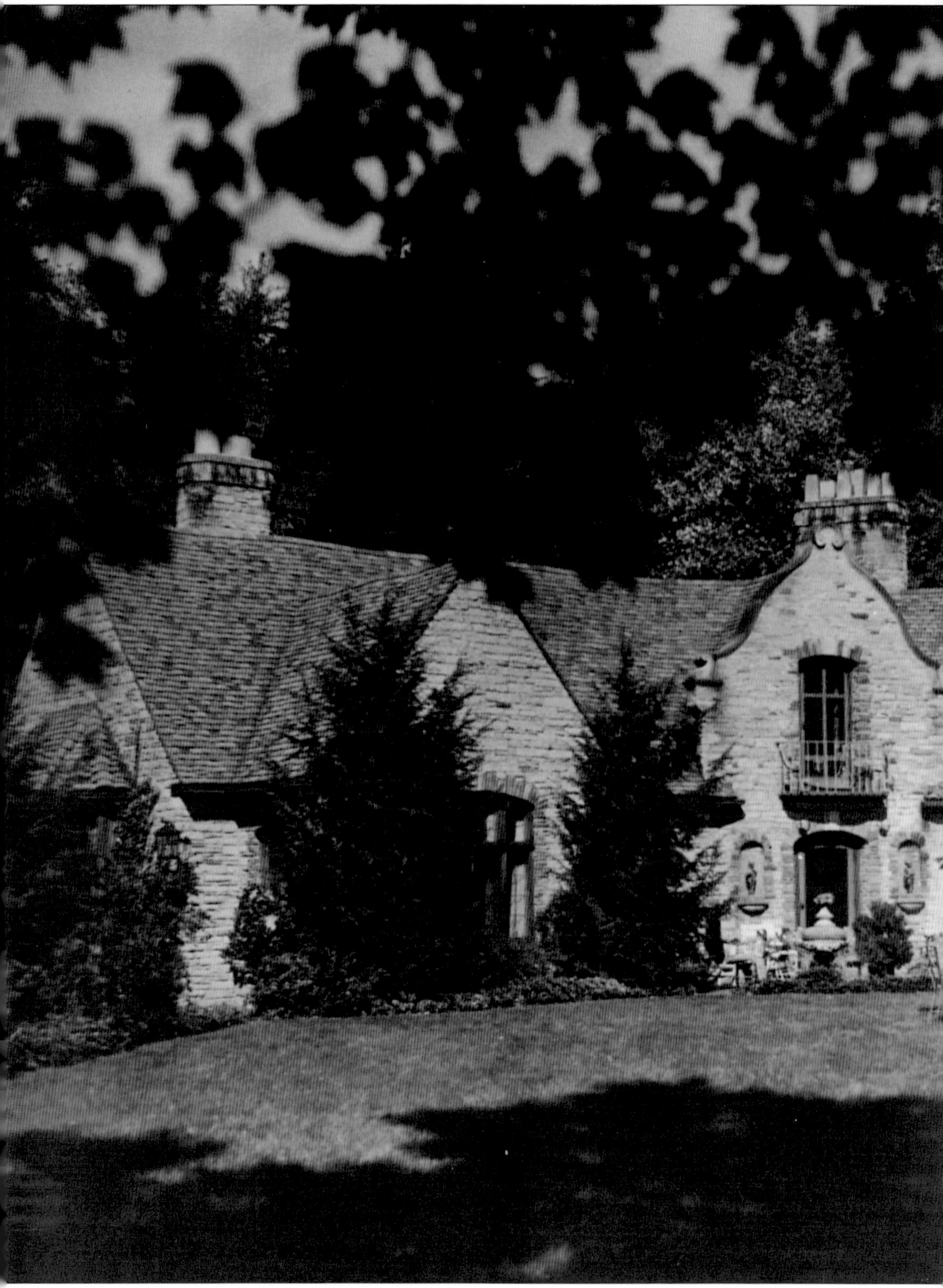

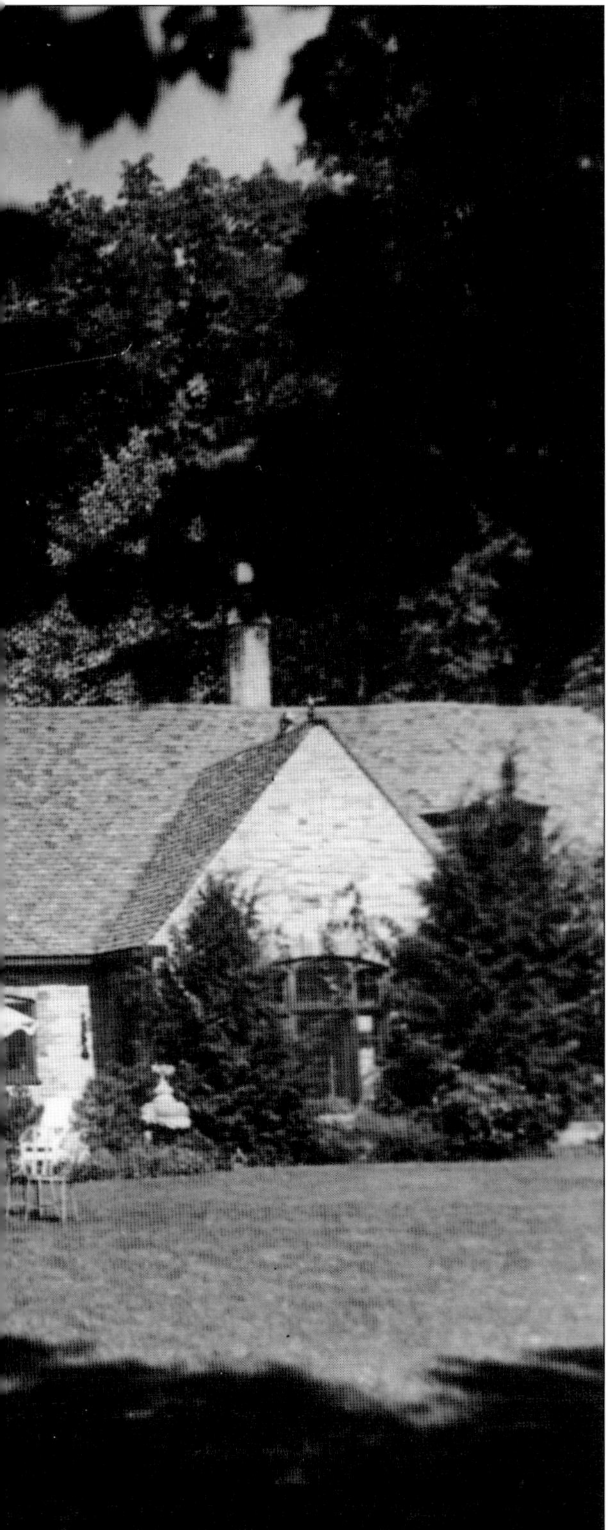

The Squires' Country Home. Designed by the noted architectural firm of Walker and Weeks, the Manor House, as it has been known for decades, was built as a surprise for Eleanor Squire in 1930 while she was abroad in Europe. On the transom above its doorway and carved in stone reads: "Parva Domus, Magna Quies." This translates to "Small House, Great Quiet." The Manor House was completed when Andrew Squire was 80 years old, and he would only get to enjoy this summer retreat for four years before he passed away in 1934. Eleanor Squire would only see seven years there before her death in 1937.

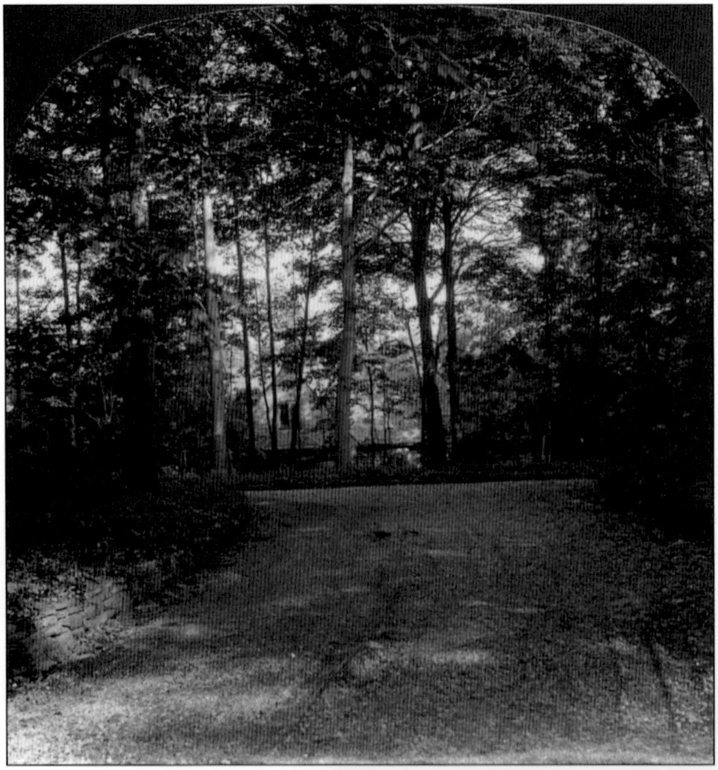

The Driveway to the Manor House. This view of the Manor House from the driveway approach is looking southward to the rear of the house. This portion of the drive still exists but is no longer accessible by vehicle from Fairmount Boulevard. It is now a part of the many walking trails that weave their way around the Manor House and the farm as a whole. Note the low sandstone wall bordering the driveway. These rocks can still be seen, though most are somewhat buried and overgrown.

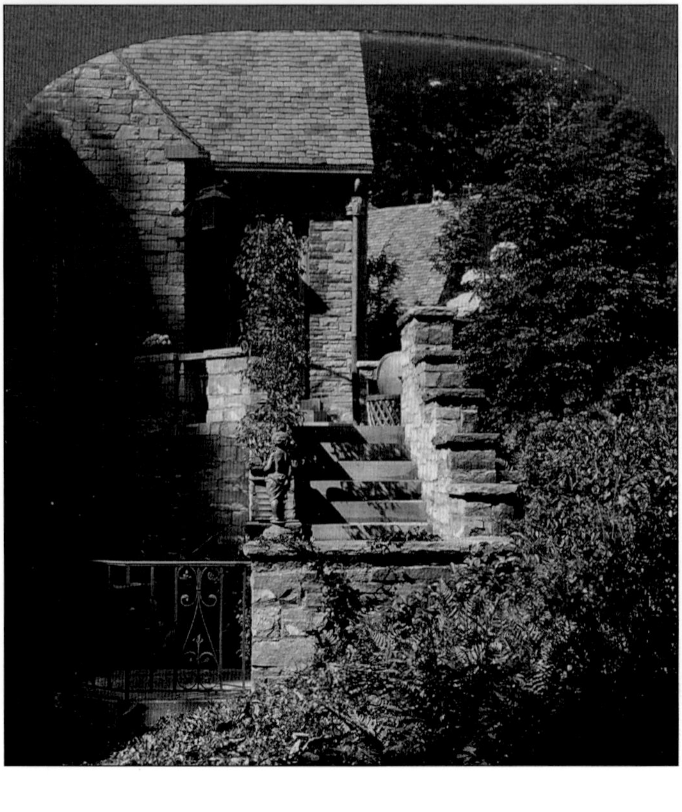

View of the Manor House, Looking North. This c. 1930 photograph shows the Manor House in summertime. All around the foundation and throughout the gardens surrounding the Manor House were dozens of elaborate fountains and statues, such as those visible here. Weather and the ravages of time have taken their toll on many of these stone beauties, but the majority of them still exist and can be seen today in the atrium of the Manor House and throughout the gardens.

A Gazebo and Spring Flowers. Nestled amidst the arboretum and naturally forested areas throughout the Squires' property were scenes such as this. Pictured is a natural path that curves through patches of spring-blooming bulbs, including daffodils, leading to a quaint gazebo where one can sit and enjoy the view. Tulips do not survive at Valleevue Farm due to the heavy deer population.

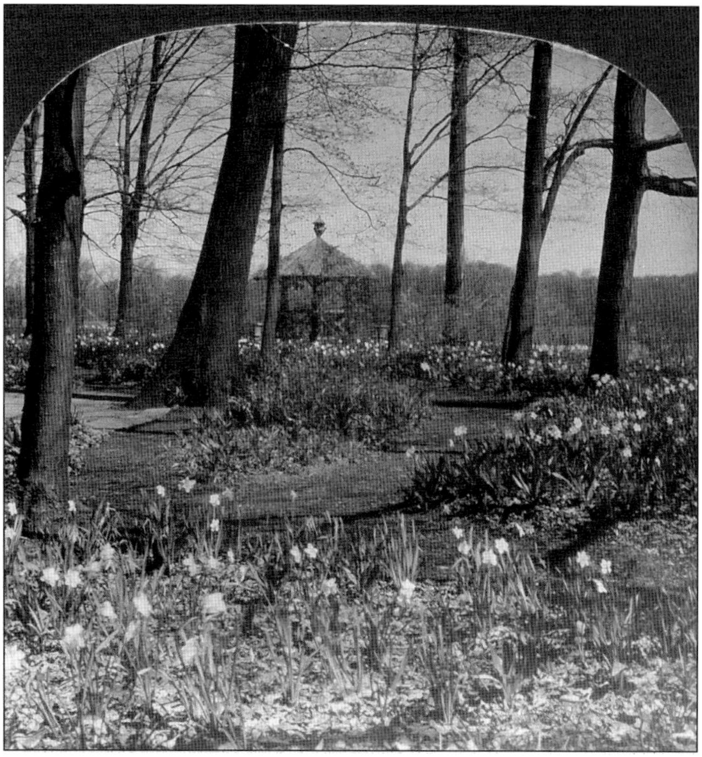

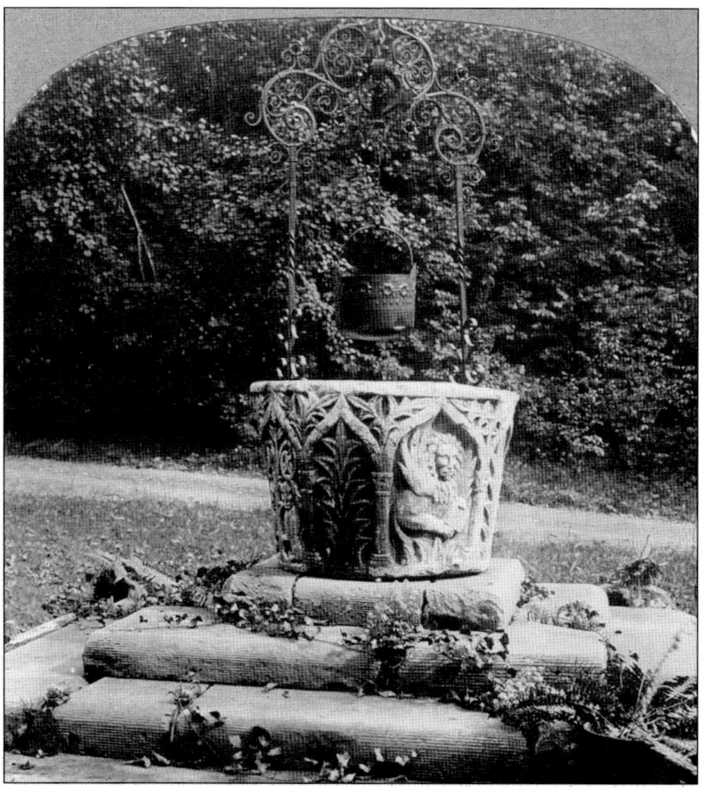

The Wishing Well. This 1925 photograph includes the wishing well perched atop its pedestal. It looks as though the fern in the lower right corner could be destined for planting nearby. Before ever living there, the Squires often used their property as a place to contemplate nature and recreate. Several dated photographs, including this one, were taken many years before the Manor House was built and the Squires' subsequent occupation of that residence.

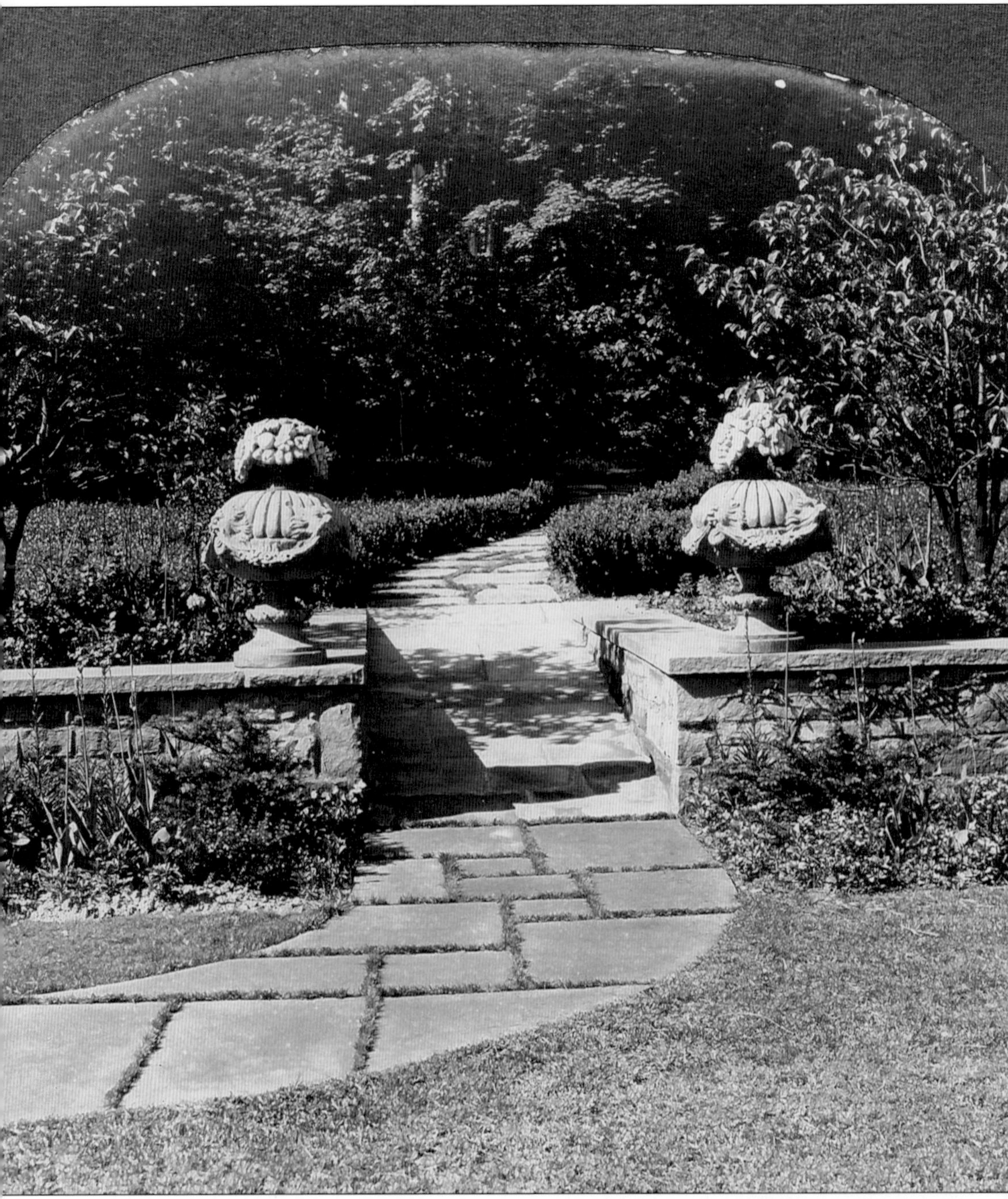

Stereoscope of Garden. By all accounts, gardening was not only Eleanor Squire's hobby but verily her passion. Mrs. Squire (widow of Sidney Guy Sea of Chicago, Illinois) would be instrumental in garnering public support and interest in botany and gardening and would be a key figure in the founding of the Garden Center of Greater Cleveland, which would become The Cleveland Botanical Garden. The resource library at the Cleveland Botanical Garden, The Eleanor Squire Library, is named in honor of all her contributions to gardening in the Greater Cleveland community.

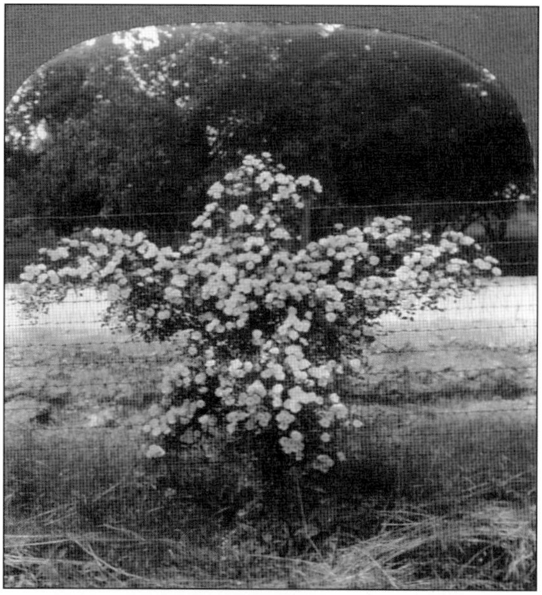

Flowering Plants. This pair of 1928 photographs shows some of the cultivated flora that was around the property. Pictured are a cluster of Asiatic lilies (below) and a multiflora rose (right) climbing along a section of fence. Multiflora roses like the one shown can still be found in abundance all over the property today.

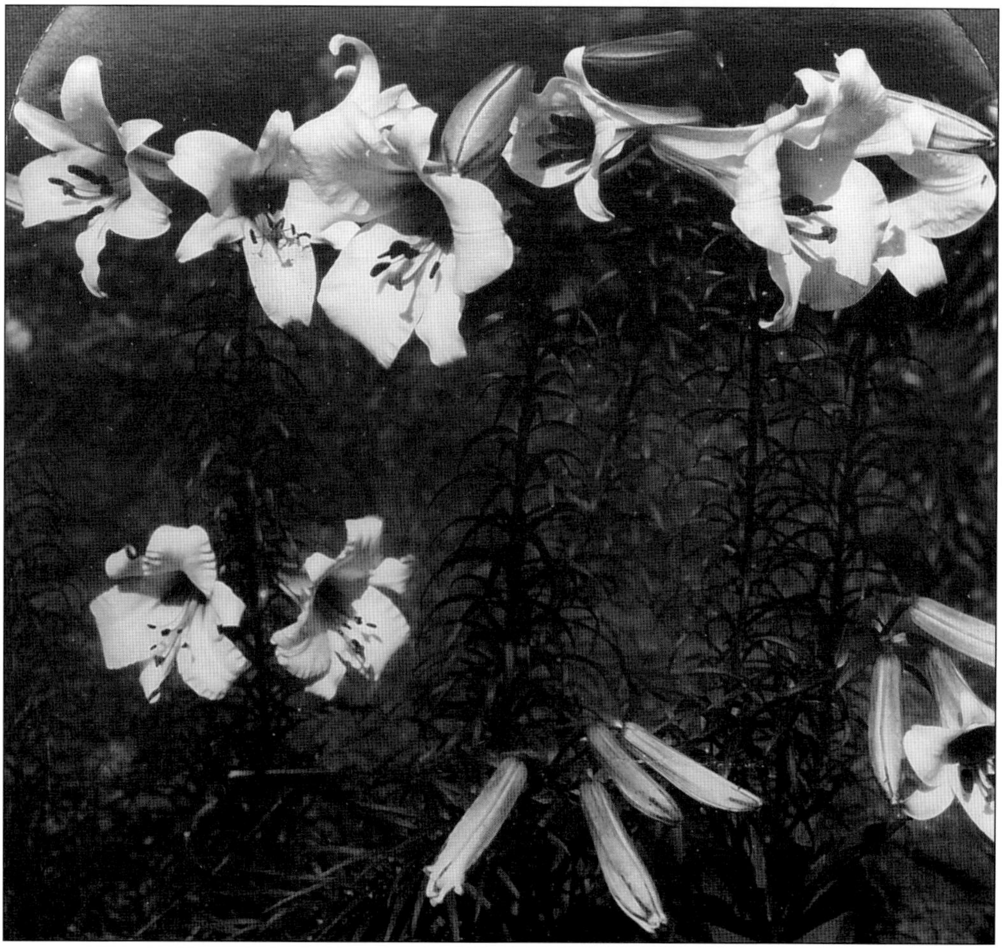

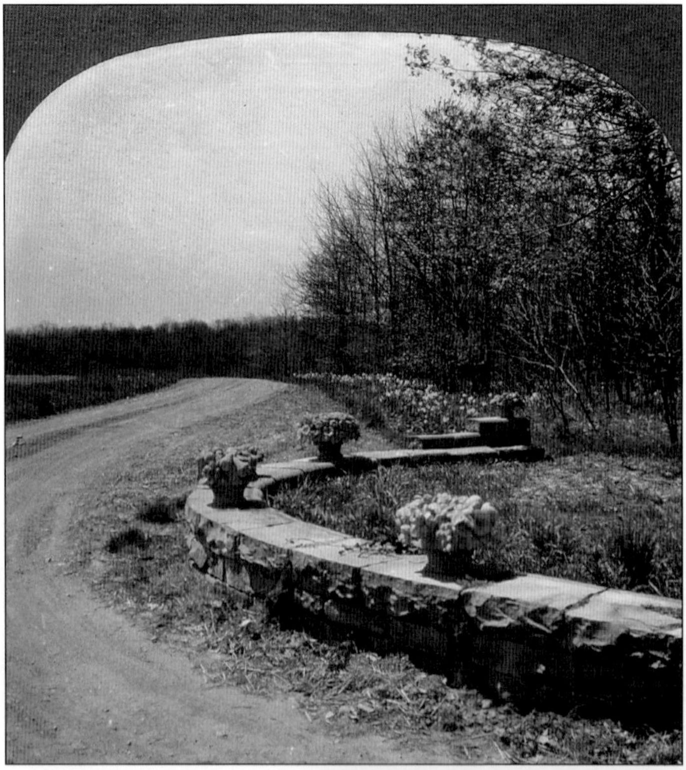

Stone Circle. This stone circle is still situated where it was in this 1925 photograph. Though the plants and the type of wall stone have changed since Andrew Squire's time, there is now a bust of him mounted atop a pillar in the center of this circle overlooking his "vallee vue."

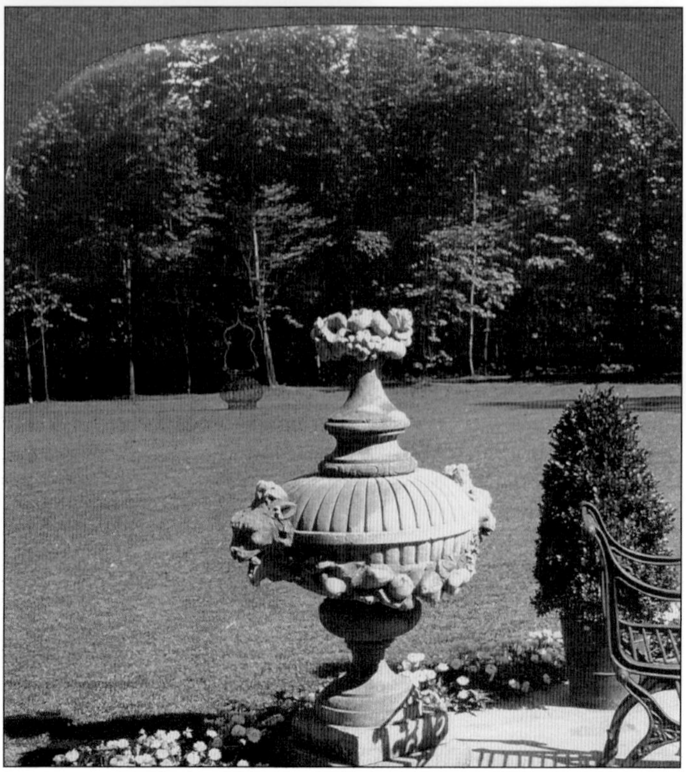

Ornamentation of the Lawn. Depicted in this c. 1931 photograph is one of the many and varied examples of the ornamentation and adornments found throughout the lawn and foundation areas of the Manor House during the Squire's residency.

EXTRACT NO. 1

This is an extract from Mr. Squire's original Will of February 12, 1919, conveying the central 150 acres to the University on his death.

Subject to these bequests and provisions for my wife, I give, devise and bequeath to the Western Reserve University of Cleveland, Ohio, for the benefit of the Women's College, to be turned over to the Western Reserve University, at such convenient time as the Western Reserve University may request, by the Citizens Savings Trust Company, after my decease, my farm in Orange Township, Cuyahoga County, Ohio, consisting of about one hundred fifty (150) acres; ***it being my especial idea that this farm shall be held in perpetuity for the use and benefit of the teachers and students of the Women's College of the Western Reserve University,*** I desire it cultivated and preserved as a farm for educational purposes, and to be a place where the practical duties of life may be taught; where the teachers and students can come in close contact with Mother Earth; and where those needing rest and recreation may obtain it. ***I desire that the woods thereon be kept and improved, so that all the young women in any way connected with the Western Reserve University may learn practical botany, may learn to love and enjoy the beauties of nature, trees and flowers; and obtain a breadth of view of life that can only come from a familiarity with nature.*** I desire that so far as possible the products of the farm be used in connection with the Women's College; and that, if my wife joins me in this trust, this bequest be known as The Eleanor and Andrew Squire Endowment for the Women's College of Western Reserve University. It is my hope that our estates will aggregate sufficient to more than permanently endow and support the farm. ***If the expenses of the farm are less that the income from my or our estates, then any other and further income, which should in time be considerable, from our estates, to be expended by Western Reserve University in erecting a suitable Teachers' and Students' Home on said farm, - if I have not already built on, - and in forwarding the educational work of the Women's College in such other manner as the Trustees' or Executive Committee of the University shall determine.***

ANDREW SQUIRE'S WILL. Andrew Squire's gift to the Western Reserve University ensured that students from all different backgrounds would have a place to interact with all the many elements and facets of nature. His will states: "I desire it cultivated and preserved as a farm for educational purposes, and to be a place where the practical duties of life may be taught; where the teachers and students can come in close contact with Mother Earth." Since the property transferred in 1937, exactly that has happened and will continue to happen in perpetuity, thanks to his generosity.

Western Reserve University is to become the principal beneficiary of the estate of the late Andrew Squire, distinguished lawyer, who died Jan. , according to Mr. Squire's will filed late yesterday for probate.

Making provision for her lifetime of his widow, Mrs. Eleanor Seymour Squire and for his son and daughter-in-law, Mr. and Mrs. Carl A. Squire; bequeathing $10,000 to the endowment fund of Trinity Cathedral and smaller sums to three servants "for faithful services", the will reveals that Reserve eventually is to receive and hold in perpetuity for its "Women's College" (Flora Stone Mather) the Squire "Valleyvue Farm" of 220 acres in Orange township for education purposes.

TRANSFER MEMO. This 1934 memo states, "Western Reserve University is to become the principal beneficiary of the estate of the late Andrew Squire, distinguished lawyer . . . [and that Western Reserve University] is to receive and hold in perpetuity for its 'Women's College' (Flora Stone Mather) the Squire 'Valleyvue [*sic*] Farm.' Mr. Squire's will is a reaffirmation of the faith in the process of higher education that actuated so many of the pioneers of the Western Reserve in founding and their successors in supporting so many educational institutions."

Two

The Flora Stone Mather College Years 1938–1972

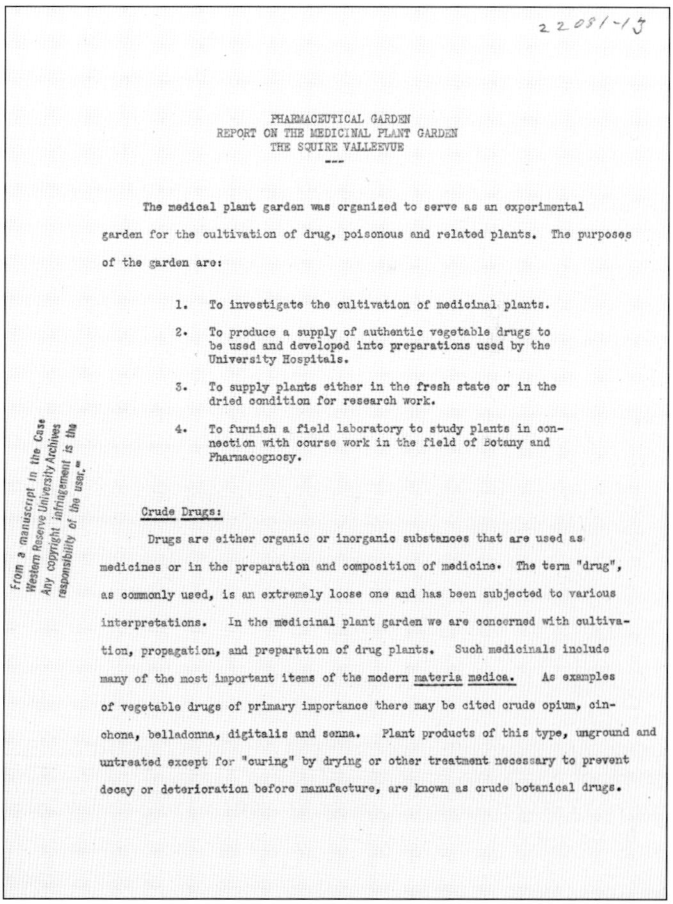

PHARMACEUTICAL GARDEN
REPORT ON THE MEDICINAL PLANT GARDEN
THE SQUIRE VALLEEVUE

The medical plant garden was organized to serve as an experimental garden for the cultivation of drug, poisonous and related plants. The purposes of the garden are:

1. To investigate the cultivation of medicinal plants.
2. To produce a supply of authentic vegetable drugs to be used and developed into preparations used by the University Hospitals.
3. To supply plants either in the fresh state or in the dried condition for research work.
4. To furnish a field laboratory to study plants in connection with course work in the field of Botany and Pharmacognosy.

Crude Drugs:

Drugs are either organic or inorganic substances that are used as medicines or in the preparation and composition of medicine. The term "drug", as commonly used, is an extremely loose one and has been subjected to various interpretations. In the medicinal plant garden we are concerned with cultivation, propagation, and preparation of drug plants. Such medicinals include many of the most important items of the modern materia medica. As examples of vegetable drugs of primary importance there may be cited crude opium, cinchona, belladonna, digitalis and senna. Plant products of this type, unground and untreated except for "curing" by drying or other treatment necessary to prevent decay or deterioration before manufacture, are known as crude botanical drugs.

Report on Pharmaceutical Garden. Upon the transfer of Squire Valleevue Farm to Western Reserve University, several assessments and reports were made of its various attributes and inventories. Here is part of a report on the pharmaceutical garden. It states, "The medical plant garden was organized to serve as an experimental garden for the cultivation of drug, poisonous and related plants."

PHILIP PUSERER
GILBERT LEPPELMEIER
HARRY CLARK

YELLOWSTONE 0756

ETHELWYN HARRISON
AND
JOHN ROYS PETERSON
LANDSCAPE DESIGNERS
2460 FAIRMOUNT BOULEVARD
CLEVELAND, OHIO

February 15, 1934

Mr. Sidney Wilson
Western Reserve University
Cleveland, Ohio

Dear Mr. Wilson:

With this letter I am enclosing my suggestion for the future developments of Mr. Squire's farm.

For the time being, to work toward the enclosed project (Re: Sustenance Farming) I would suggest that the surplus of milk and cream and poultry products, be used in the Hospital and cafeteria; that such vegetables be grown as are desired by these institutions for their use. Many of the other products such as fruit, honey and syrup, could be used by the University.

You might be interested in knowing that the Dairy is not only serving its purpose as a milk producing unit, but also that at least eight young bulls and cows have been sold recently to farmers in the surrounding counties to build up their local herds. This was one aim that Mr. Squire was interested in; the betterment of the herds of the average farmer so that the consumer would receive higher quality milk.

There is much information with map and records which I feel you might be interested in inspecting. If at any time you care to come to my office, I will be very glad to show you all I have.

Very sincerely yours,

EH/MB
Encl.

Report by Violet Ethelwyn Harrison. Ethelwyn Harrison did not stay in the employ of the farm after the transfer. She would go back to her private landscape architecture practice and stay in Northeast Ohio her entire life. Before leaving, she offered this report with her suggestions to the university on its use of Valleevue Farm. It was titled "Squire Valleevue as a Laboratory to Further the Art of Living."

YELLOWSTONE 0756

ETHELWYN HARRISON
LANDSCAPE ARCHITECT
2460 FAIRMOUNT BOULEVARD
CLEVELAND, OHIO

SQUIRE VALLEEVUE AS A LABORATORY TO FURTHER THE ART OF LIVING

I believe that the Squire Valleevue Farm can be developed in furtherance of Mr. Squire's ideals and desires and a laboratory to further the real art of living.

In view of the probable increase in leisure due to shorter hours of work, and the likelihood of generally restricted incomes, there appears to be a social obligation to encourage young people to have not only a home of their own but a home site--not of a mere city lot--but from one to five acres of land. If the foregoing view is accepted, there should be available a thorough but not necessarily a formal training in how to obtain the utmost pleasure and profit out of home and grounds. This should stimulate a genuine and discriminating love of Nature. This plan has particularly in view the so-called "White Collar" classes in a form of what is now known as "Sustenance" living or farming.

There should be actual examples on the farm of such demonstration places. The gardener's house on S.O.M. Center Road and the dairyman's place on North Woodland Road could be very readily adapted as demonstration plots.

With the place as a practical laboratory, there should be training offered not in professional Landscape Architecture or Horticulture courses but designed rather to develop and train in a practical manner a real love of nature and its utilization in a home in the country without a large amount of money. There should be taught, first of all, the fundamentals of good design; the characteristics of different types of soil; the common ordinary things both for their beauty and utility. There should be instruction in growing beans, peas and corn; with economic facts such as that the average family needs six to seven hundred pounds of vegetables every year; that so-many feet of beans, peas or corn will generally produce enough for the average family.

There is any amount of pleasure and real art in living on a place in the country. The United States is over its rough pioneering stage, consequently the home owner should learn more of the art and proper use of leisure; and to realize that there is no disgrace in knowing how to properly hoe or use a mattock; that an attractive garden can be had with very little expense if properly arranged and taken care of; that it is the height of folly to try to grow sun-loving perennials in a wooded area or shady spot. They should learn to know that a rock garden is not a pile of rocks in the center of a lawn; that it is better to grow a ground cover in the shade than to use grass; that the majority of bulbs will do as well naturalized in a wild garden without care as the fine hybrids used in formal gardening with skilled care. They should learn to know that the Flowering Dogwood, the Wild Crab and other similar types of plants are as fine as any exotic that can be grown in this climate; that there is just as much real beauty in a spike of well-grown Hollyhock or a carrot top as in a French Hybrid Peony.

Suggested Uses for May Squire House. This 1934 report makes several recommendations for the potential uses of what would become the May Squire House. It focuses on "establishing a group of courses as a coordinated unit with a view to preparing college women to be the kind of homemakers [that] modern society needs and expects to find in the college graduate."

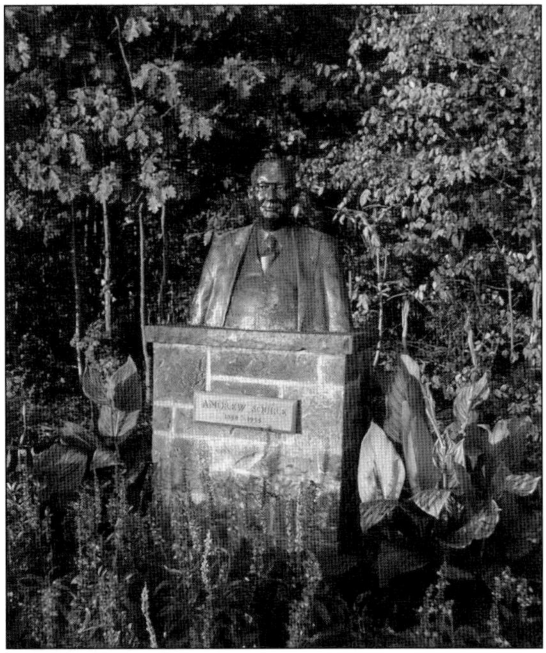

Bust of Andrew Squire. This statue of Andrew Squire was erected during the Flora Stone Mather College years. Pictured amongst cannas and salvia, it is sited at the peak of the property looking east over the Chagrin Valley and enjoys some of the most impressive views of the property. It can be seen en route to the Manor House. (Andrew Toepfer.)

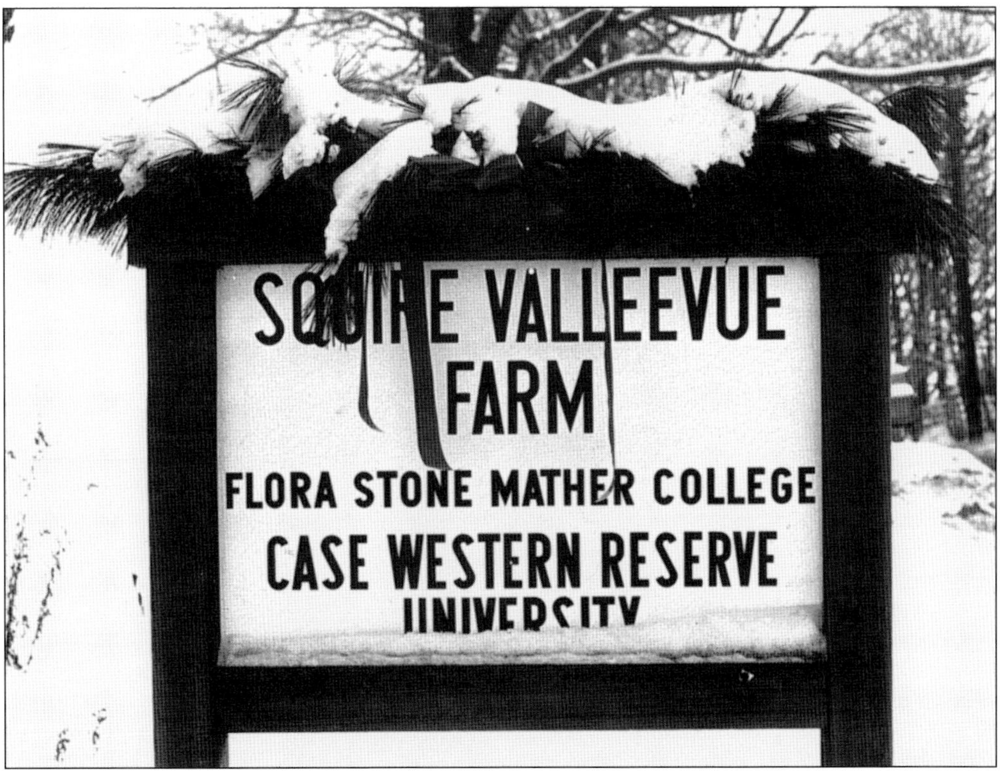

Squire Valleevue Farm Sign. Pictured is one of the original entrance signs placed at the entrance on Fairmount Boulevard. It has since been replaced with hand-carved redwood signs that are also used throughout the property to indicate the locations of the other buildings and sites. (Andrew Toepfer.)

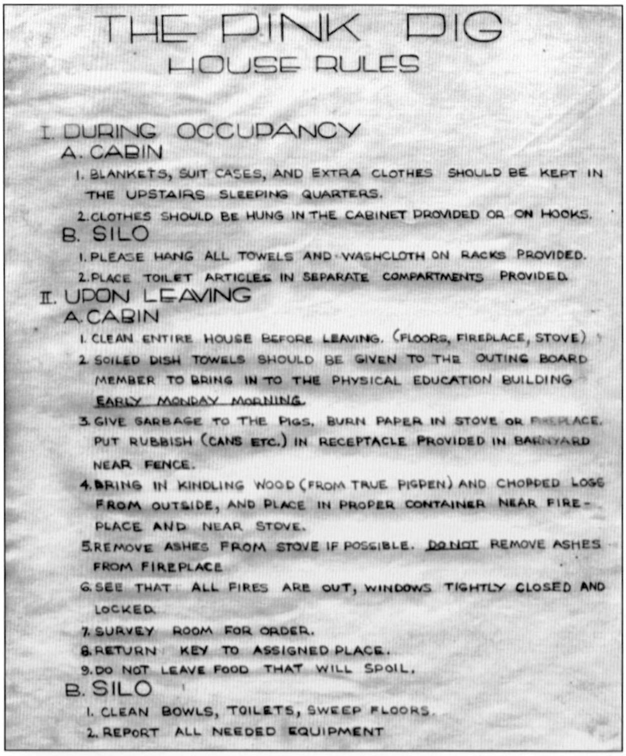

THE PINK PIG
HOUSE RULES

I. DURING OCCUPANCY
A. CABIN
1. BLANKETS, SUIT CASES, AND EXTRA CLOTHES SHOULD BE KEPT IN THE UPSTAIRS SLEEPING QUARTERS.
2. CLOTHES SHOULD BE HUNG IN THE CABINET PROVIDED OR ON HOOKS.
B. SILO
1. PLEASE HANG ALL TOWELS AND WASHCLOTH ON RACKS PROVIDED.
2. PLACE TOILET ARTICLES IN SEPARATE COMPARTMENTS PROVIDED.
II. UPON LEAVING
A. CABIN
1. CLEAN ENTIRE HOUSE BEFORE LEAVING. (FLOORS, FIREPLACE, STOVE)
2. SOILED DISH TOWELS SHOULD BE GIVEN TO THE OUTING BOARD MEMBER TO BRING IN TO THE PHYSICAL EDUCATION BUILDING EARLY MONDAY MORNING.
3. GIVE GARBAGE TO THE PIGS. BURN PAPER IN STOVE OR FIREPLACE. PUT RUBBISH (CANS ETC.) IN RECEPTACLE PROVIDED IN BARNYARD NEAR FENCE.
4. BRING IN KINDLING WOOD (FROM TRUE PIGPEN) AND CHOPPED LOGS FROM OUTSIDE, AND PLACE IN PROPER CONTAINER NEAR FIRE-PLACE AND NEAR STOVE.
5. REMOVE ASHES FROM STOVE IF POSSIBLE. DO NOT REMOVE ASHES FROM FIREPLACE
6. SEE THAT ALL FIRES ARE OUT, WINDOWS TIGHTLY CLOSED AND LOCKED.
7. SURVEY ROOM FOR ORDER.
8. RETURN KEY TO ASSIGNED PLACE.
9. DO NOT LEAVE FOOD THAT WILL SPOIL.
B. SILO
1. CLEAN BOWLS, TOILETS, SWEEP FLOORS.
2. REPORT ALL NEEDED EQUIPMENT

THE RULES. These were the simple rules posted for the girls at Flora Stone Mather College that they were asked to obey while at the farm. George and Hattie Krummers were Squire Valleevue Farm's superintendents during this era.

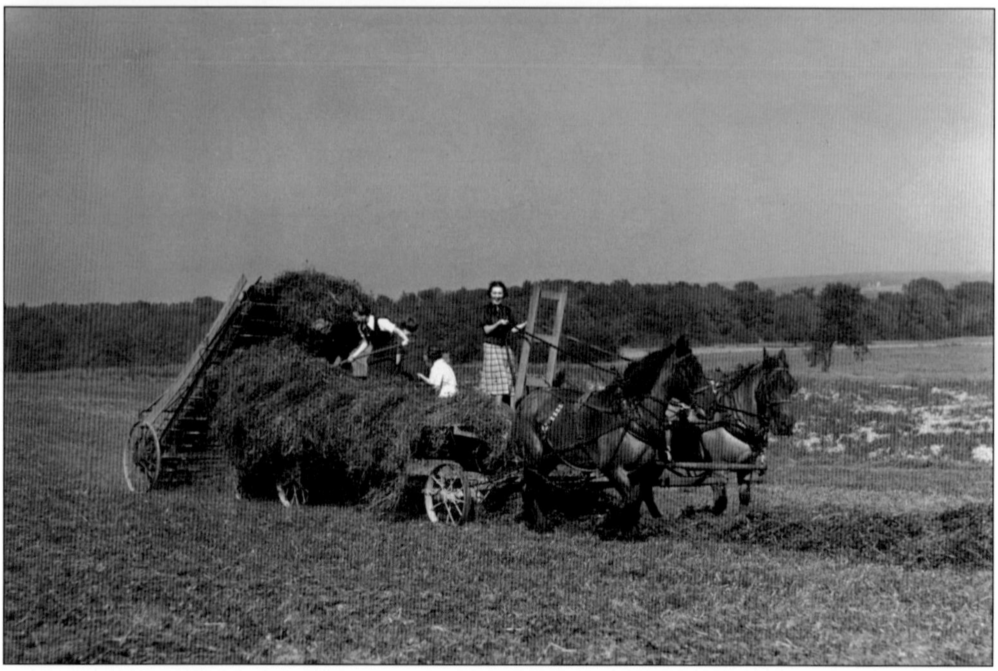

YOUNG WOMEN DRIVING WORK HORSES. Transferring ownership of Squire's Valleevue Farm did not negate the massive amount of chores that had to be undertaken on a working farm. Here, instead of the farmhands who worked for Andrew Squire, Flora Stone Mather College students mount the tasks of harvest and collecting hay.

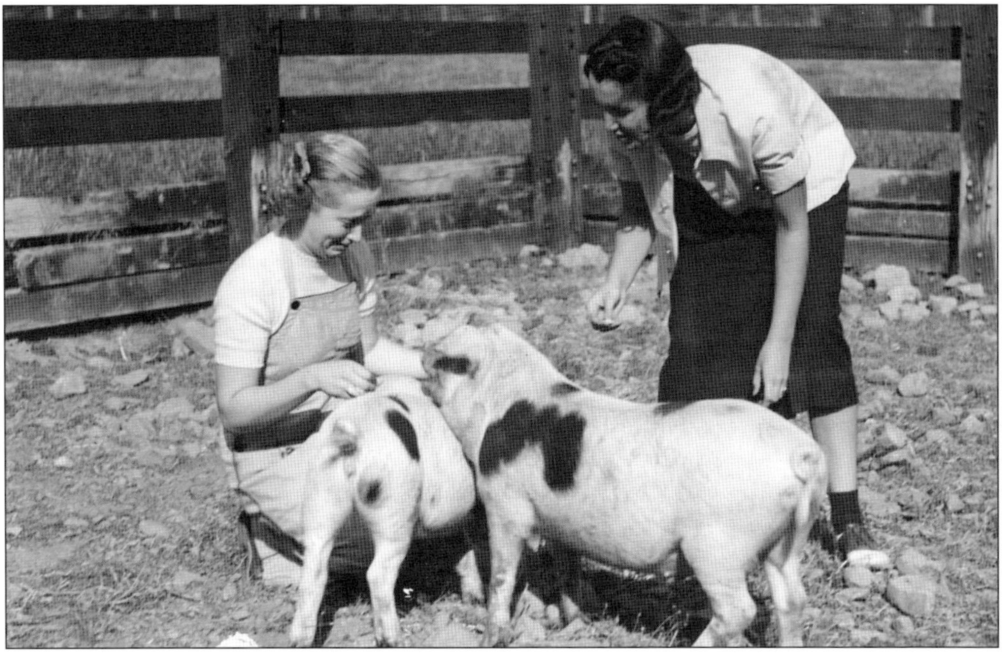

Young Women with Pigs. It was Andrew Squire's sincerest wish that after his death the young women of the Western Reserve University would have a place to enjoy the many wonders of nature, including animal husbandry. Here, two young ladies are getting a very close introduction to some real "pink pigs."

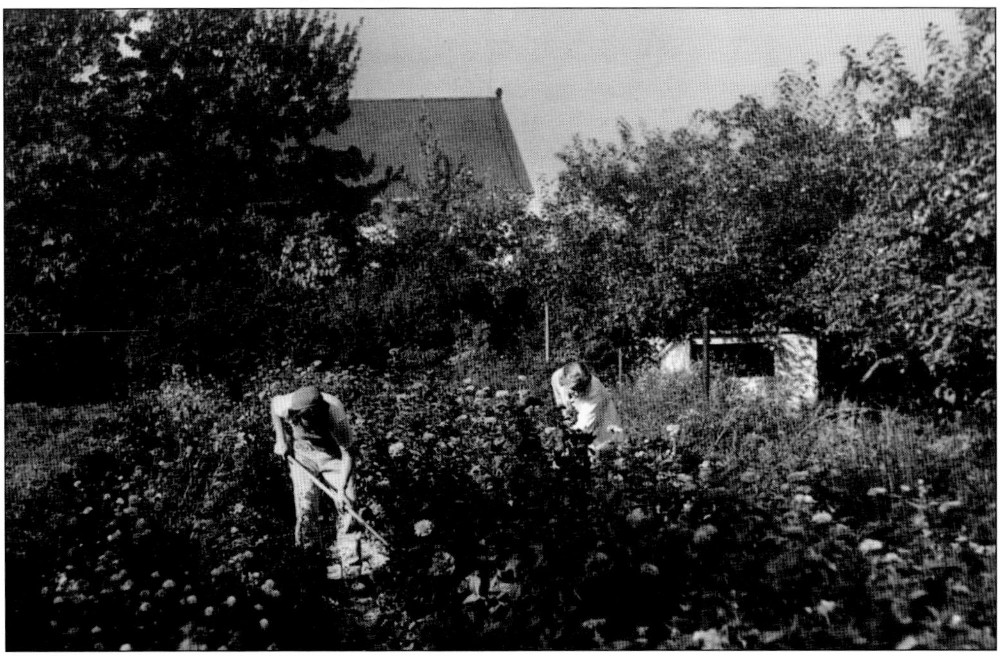

Two Students Working in the Pharmaceutical Garden. In this undated photograph, two girls pull weeds among the rows in the Pharmaceutical Garden. They are working the field on the west side of the storage barn, which still exists today. This area is now part of the vast expanse of unmowed fields that are home to the farm's many wildlife species.

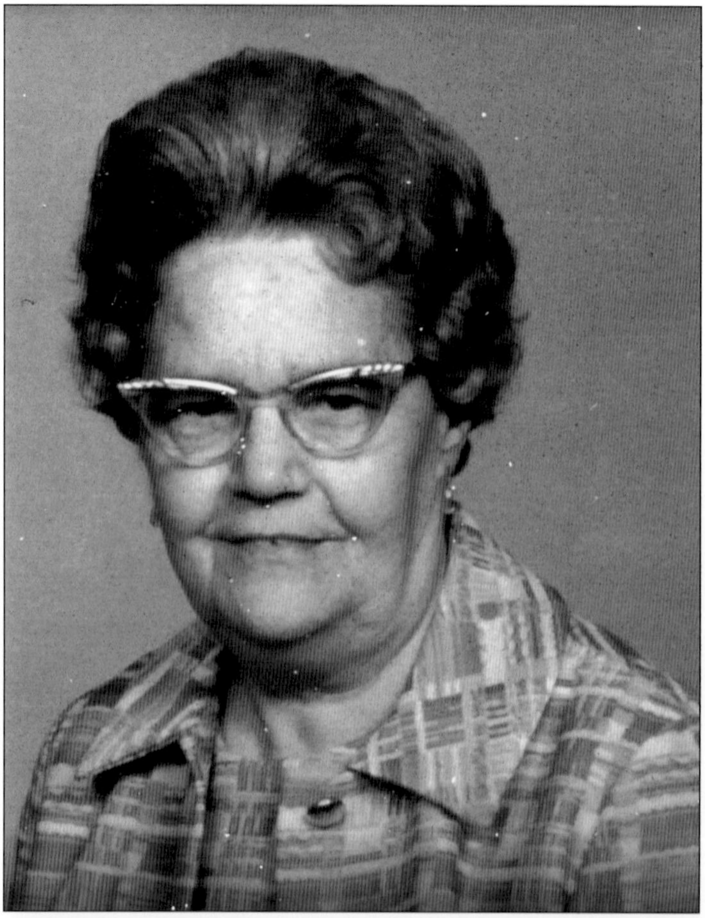

Hattie Krummers. Hattie Krummers (left) and her husband, George Krummers, were the first superintendents of Squire Valleevue Farm after it transferred to Western Reserve University. During their tenure, besides attending to the needs of the girls of Flora Stone Mather College, they grew prizewinning gladioli (below). They were planted in the fertile field formerly occupied by the pharmaceutical garden, as seen in this photograph taken by former farm director Dr. Norman Alldridge.

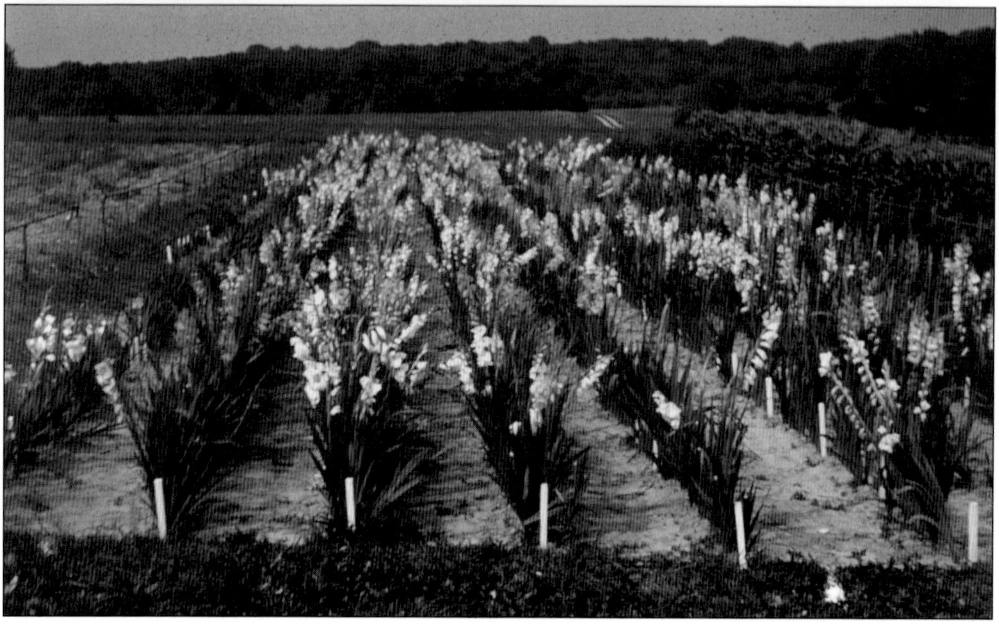

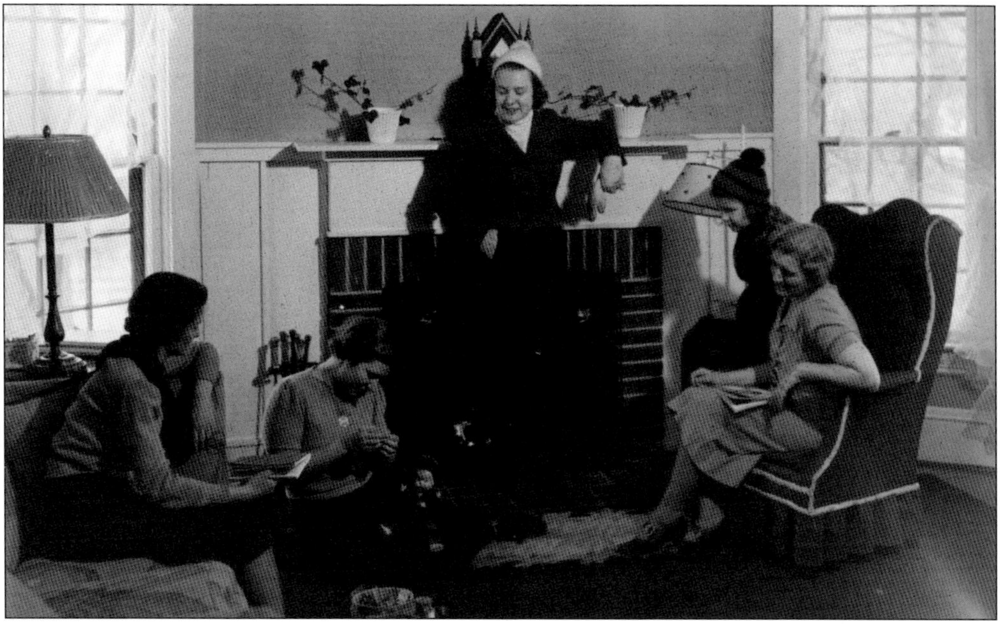

HOME ECONOMICS IN THE MAY SQUIRE HOUSE. Built in the 1850s, this farmhouse is one of the two oldest buildings on the property and was the original residence located on the dairy farm that Andrew Squire purchased. It served as a learning house and classroom where the young women would take their homemaking courses. It is the same house that is pictured in the photograph of the woman standing by the gate to the farm on page 11.

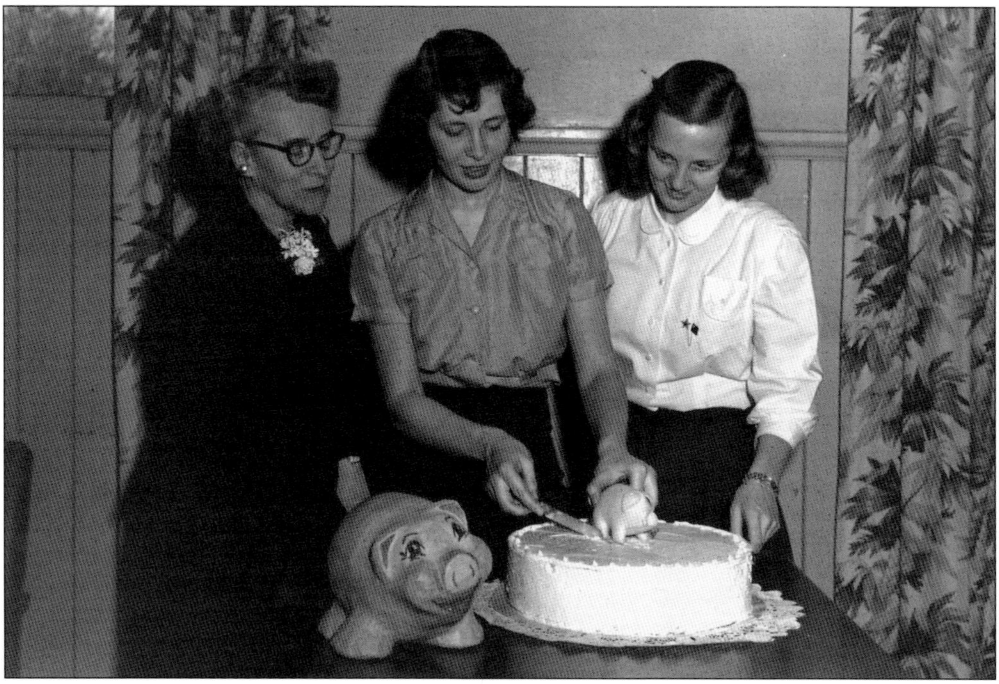

CAKE BAKING IN THE MAY SQUIRE HOUSE. Two Flora Stone Mather College girls and an instructor go over the finer points of putting the finishing touches on a pig-themed cake. The pig on the table seems to be looking on longingly for a chance to bury his snout in the frosting.

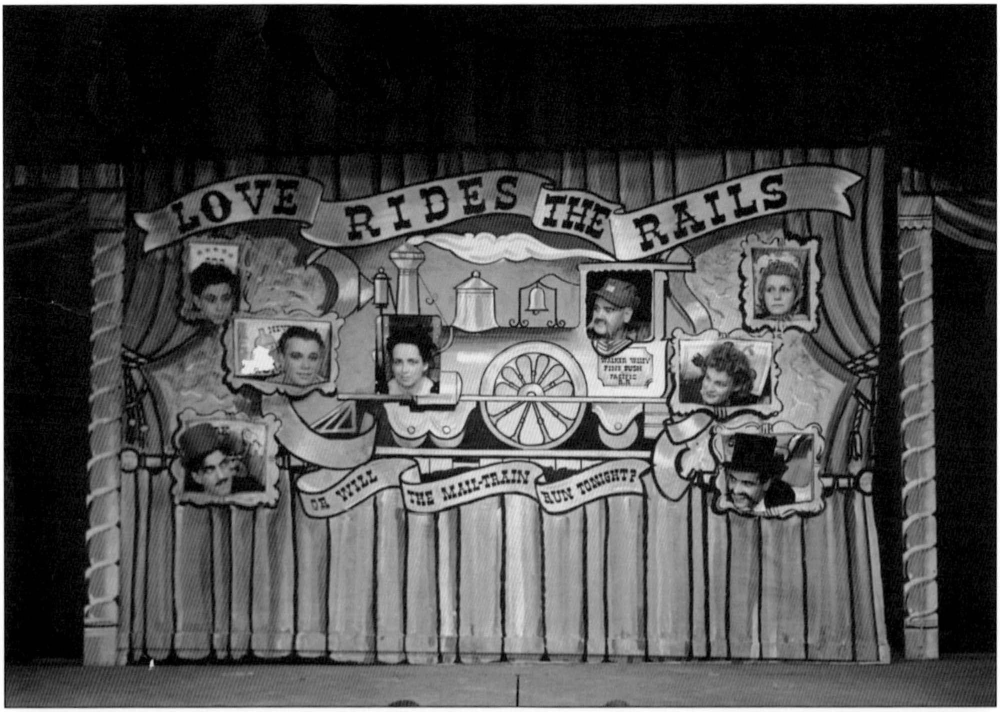

THEATER PRODUCTION IN THE MAIN BARN. After its primary use as a stable for the farm animals and before its conversion to facilities for small-animal research (such as rabbits and prairie dogs), the main barn area experienced an incarnation as a theater area for student productions of various plays and performances.

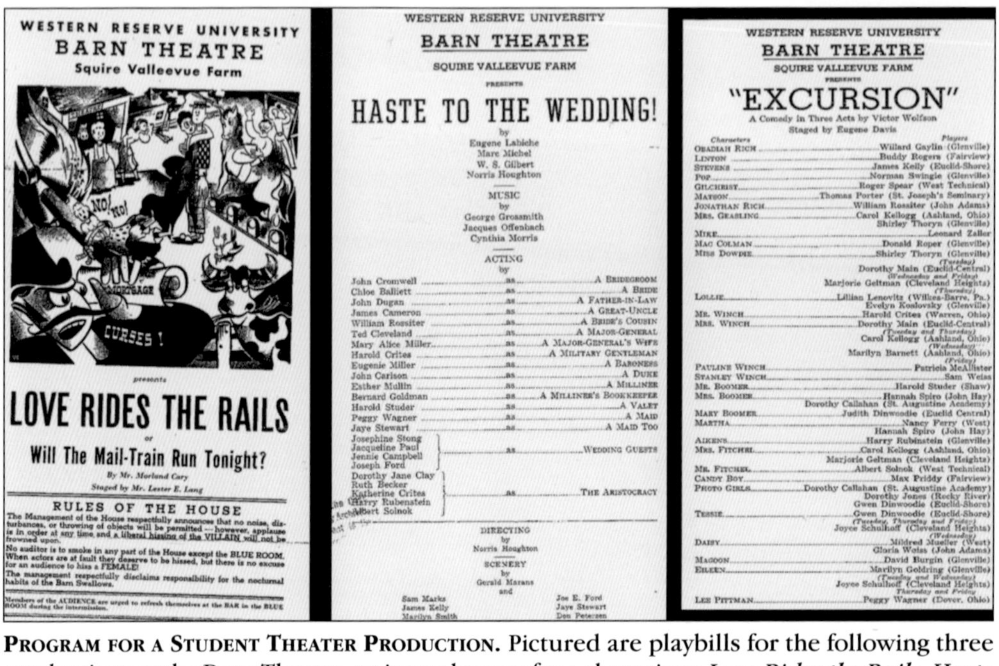

WESTERN RESERVE UNIVERSITY
BARN THEATRE
Squire Valleevue Farm

presents

LOVE RIDES THE RAILS
or
Will The Mail-Train Run Tonight?

By Mr. Morland Cary
Staged by Mr. Lester E. Lang

RULES OF THE HOUSE

The Management of the House respectfully announces that no noise, disturbances, or throwing of objects will be permitted—however, applause is in order at any time and a liberal hissing of the VILLAIN will not be frowned upon.

No auditor is to smoke in any part of the House except the BLUE ROOM. When actors are at fault they deserve to be hissed, but there is no excuse for an audience to hiss a FEMALE!

The management respectfully disclaims responsibility for the nocturnal habits of the Barn Swallows.

Members of the AUDIENCE are urged to refresh themselves at the BAR in the BLUE ROOM during the intermission.

WESTERN RESERVE UNIVERSITY
BARN THEATRE
SQUIRE VALLEEVUE FARM
PRESENTS

HASTE TO THE WEDDING!

by
Eugene Labiche
Marc Michel
W. S. Gilbert
Norris Houghton

MUSIC
by
George Grossmith
Jacques Offenbach
Cynthia Morris

ACTING
by

John Cromwell	as	A Bridegroom
Chloe Balliett	as	A Bride
John Dugan	as	A Father-in-Law
James Cameron	as	A Great-Uncle
William Rossiter	as	A Bride's Cousin
Ted Cleveland	as	A Major-General
Mary Alice Miller	as	A Major-General's Wife
Harold Crites	as	A Military Gentleman
Eugenie Miller	as	A Baroness
John Carlson	as	A Duke
Esther Mullin	as	A Milliner
Bernard Goldman	as	A Milliner's Bookkeeper
Harold Studer	as	A Valet
Peggy Wagner	as	A Maid
Jaye Stewart	as	A Maid Too
Josephine Stong, Jacqueline Paul, Jennie Campbell, Joseph Ford	as	Wedding Guests
Dorothy Jane Clay, Ruth Becker, Katherine Crites, Harry Rubenstein, Albert Solnok	as	The Aristocracy

DIRECTING
by
Norris Houghton

SCENERY
by
Gerald Marans
and

Sam Marks
James Kelly
Marilyn Smith

Joe E. Ford
Jaye Stewart
Don Petersen

WESTERN RESERVE UNIVERSITY
BARN THEATRE
SQUIRE VALLEEVUE FARM
PRESENTS

"EXCURSION"

A Comedy in Three Acts by Victor Wolfson
Staged by Eugene Davis

Characters	Players
Obadiah Rich	Willard Gaylin (Glenville)
Linton	Buddy Rogers (Fairview)
Stevens	James Kelly (Euclid-Shore)
Pop	Norman Swingle (Glenville)
Gilchrist	Roger Spear (West Technical)
Matson	Thomas Porter (St. Joseph's Seminary)
Jonathan Rich	William Rossiter (John Adams)
Mrs. Geasling	Carol Kellogg (Ashland, Ohio)
	Shirley Thoryn (Glenville)
Mike	Leonard Zaller
Mac Colman	Donald Roper (Glenville)
Miss Dowdie	Shirley Thoryn (Glenville) (Tuesday)
	Dorothy Main (Euclid-Central) (Wednesday and Friday)
	Marjorie Geltman (Cleveland Heights) (Thursday)
Lollie	Lillian Lenovitz (Wilkes-Barre, Pa.)
	Evelyn Koslovsky (Glenville)
Mr. Winch	Harold Crites (Warren, Ohio)
Mrs. Winch	Dorothy Main (Euclid-Central) (Tuesday and Thursday)
	Carol Kellogg (Ashland, Ohio) (Wednesday)
	Marilyn Barnett (Ashland, Ohio) (Friday)
Pauline Winch	Patricia McAllister
Stanley Winch	Sam Weiss
Mr. Boomer	Harold Studer (Shaw)
Mrs. Boomer	Hannah Spiro (John Hay)
	Dorothy Callahan (St. Augustine Academy)
Mary Boomer	Judith Dinwoodie (Euclid Central)
Martha	Nancy Ferry (West)
	Hannah Spiro (John Hay)
Aikens	Harry Rubinstein (Glenville)
Mrs. Fitchel	Carol Kellogg (Ashland, Ohio)
	Marjorie Geltman (Cleveland Heights)
Mr. Fitchel	Albert Solnok (West Technical)
Candy Boy	Max Priddy (Fairview)
Photo Girls	Dorothy Callahan (St. Augustine Academy)
	Dorothy Jones (Rocky River)
	Gwen Dinwoodie (Euclid-Shore)
Tessie	Gwen Dinwoodie (Euclid-Shore) (Tuesday, Thursday and Friday)
	Joyce Schulhoff (Cleveland Heights) (Wednesday)
Daisy	Mildred Mueller (West)
	Gloria Weiss (John Adams)
Magoon	David Burgin (Glenville)
Eileen	Marilyn Goldring (Glenville) (Tuesday and Wednesday)
	Joyce Schulhoff (Cleveland Heights) Thursday and Friday
Lee Pittman	Peggy Wagner (Dover, Ohio)

PROGRAM FOR A STUDENT THEATER PRODUCTION. Pictured are playbills for the following three productions at the Barn Theatre, as it was known for a short time: *Love Rides the Rails*, *Haste to the Wedding!*, and *Excursion*.

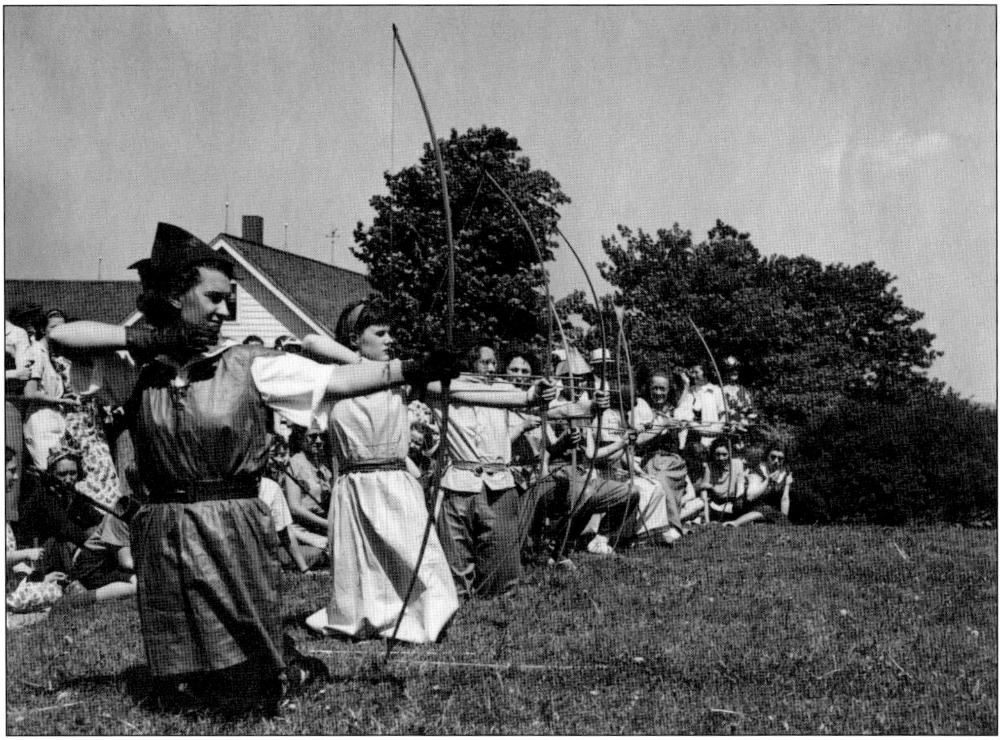

Archery. Recreational and physical educational activities were held on the farm year-round, with archery being a popular class. Here, some aspiring young Robin Hoods take aim at their target in the distance. (The costumes were presumably optional but look fun nonetheless.)

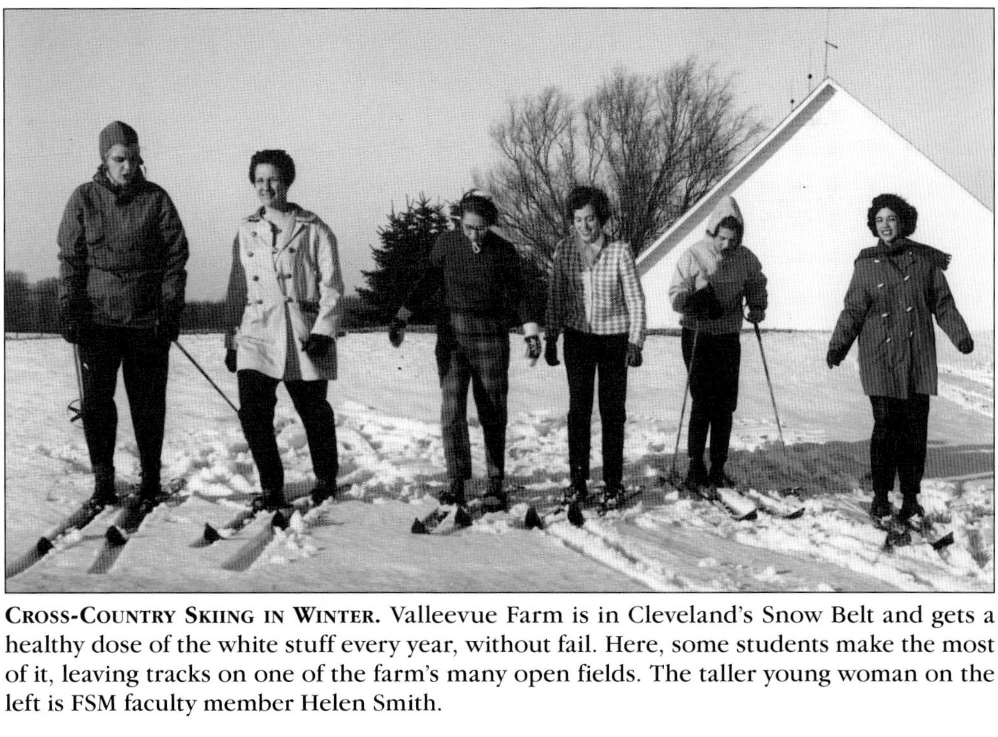

Cross-Country Skiing in Winter. Valleevue Farm is in Cleveland's Snow Belt and gets a healthy dose of the white stuff every year, without fail. Here, some students make the most of it, leaving tracks on one of the farm's many open fields. The taller young woman on the left is FSM faculty member Helen Smith.

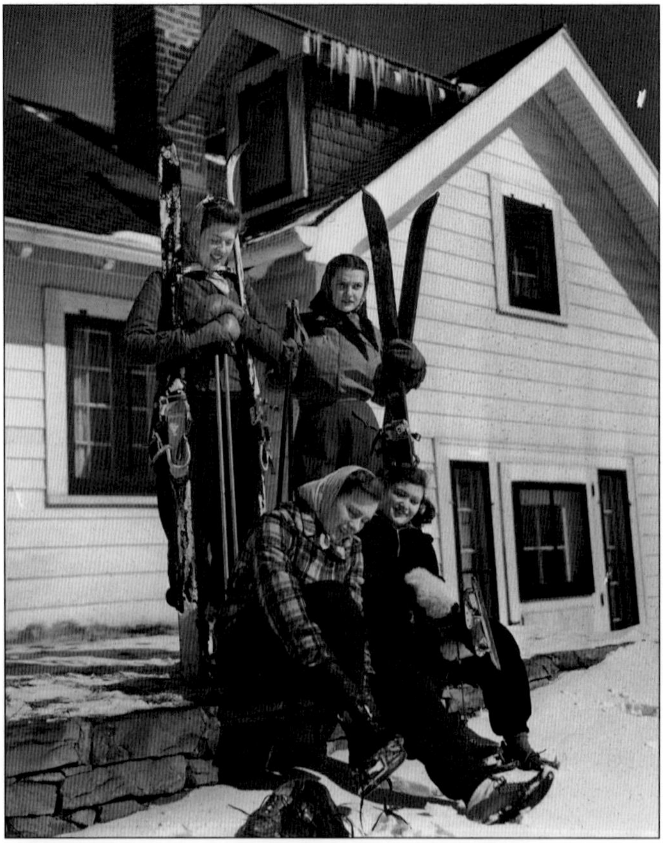

Winter Fun at Valleevue Farm. Skiing was not the only popular winter sport enjoyed by the girls and farm visitors. Within the farm's borders are several ponds that were and still are suitable for an afternoon of ice-skating (left). The many hills about the nearly 300-acre property also afforded the opportunity for some prime downhill sledding (below). Winters at University Farm today still provide ample opportunity for students, staff, faculty, alumni, and their families to engage in the many diversions that Cleveland's cold-weather months provide.

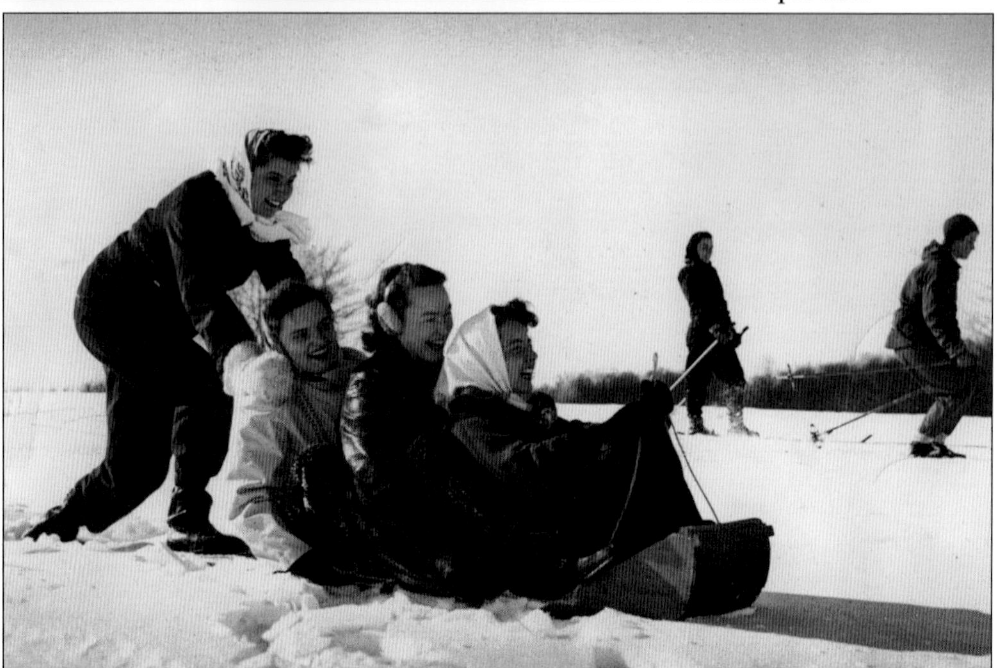

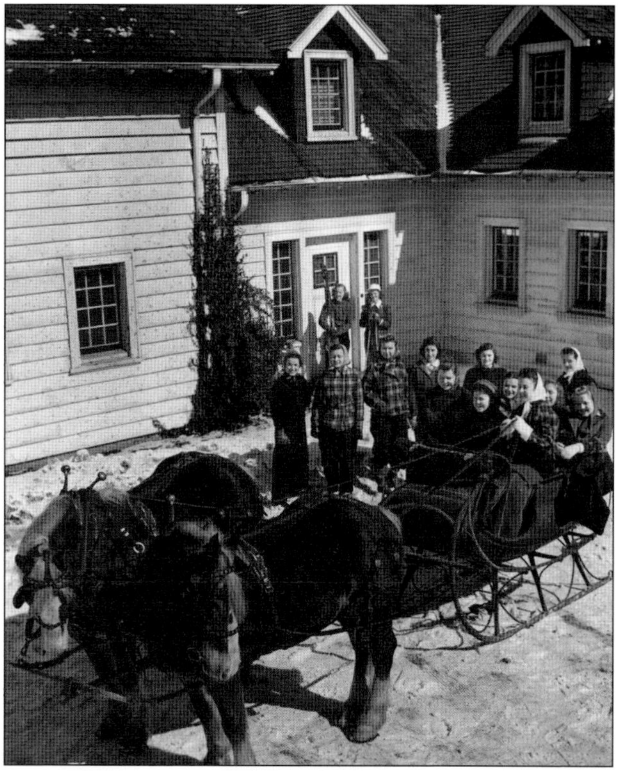

Girls with Horses and Sleigh. Another fun winter activity enjoyed by the Flora Stone Mather College women was horse-drawn sleigh rides through the fields. Here, the girls are about to embark on a trek to dash through the snow on a two-horse open sleigh. This sleigh is still around and sometimes put out around the holidays for decoration.

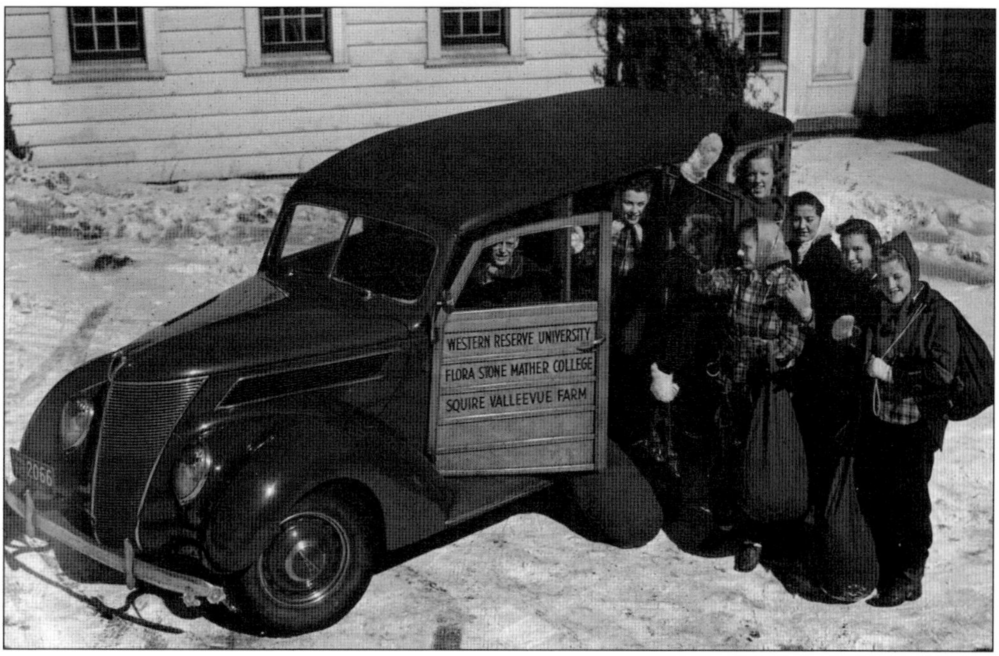

Young Ladies File into Their Station Wagon. Seen here are some Flora Stone Mather College students getting ready to depart from the farm on a winter's day. Note the inscription on the door; the vehicle was the main transportation for students between the farm and the main campus.

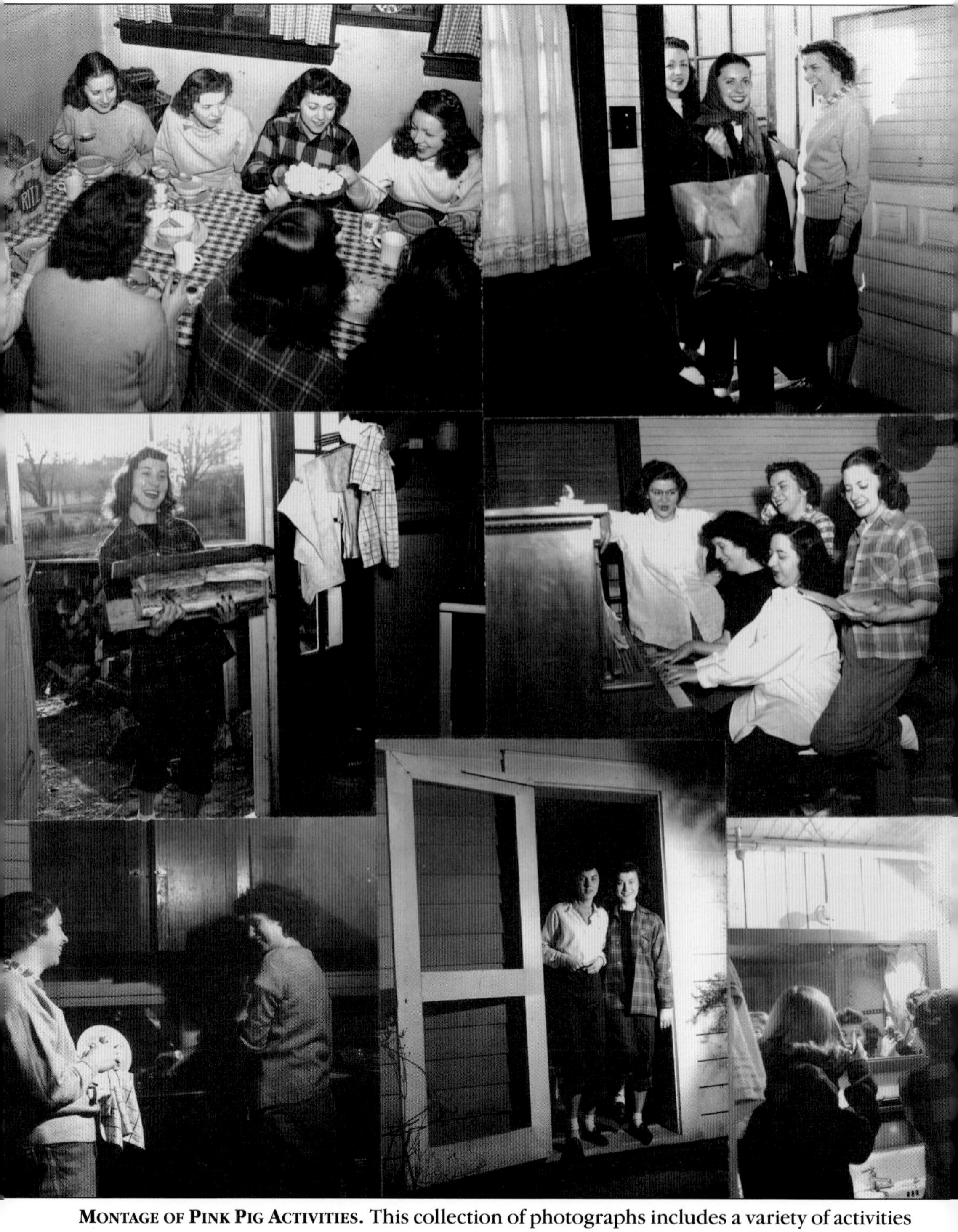

Montage of Pink Pig Activities. This collection of photographs includes a variety of activities that went on at the Pink Pig throughout the year. This beloved little cottage would be used for decades by the young ladies of Flora Stone Mather College for social, academic, athletic, and recreational activities. It is still a sought-after venue for university student groups.

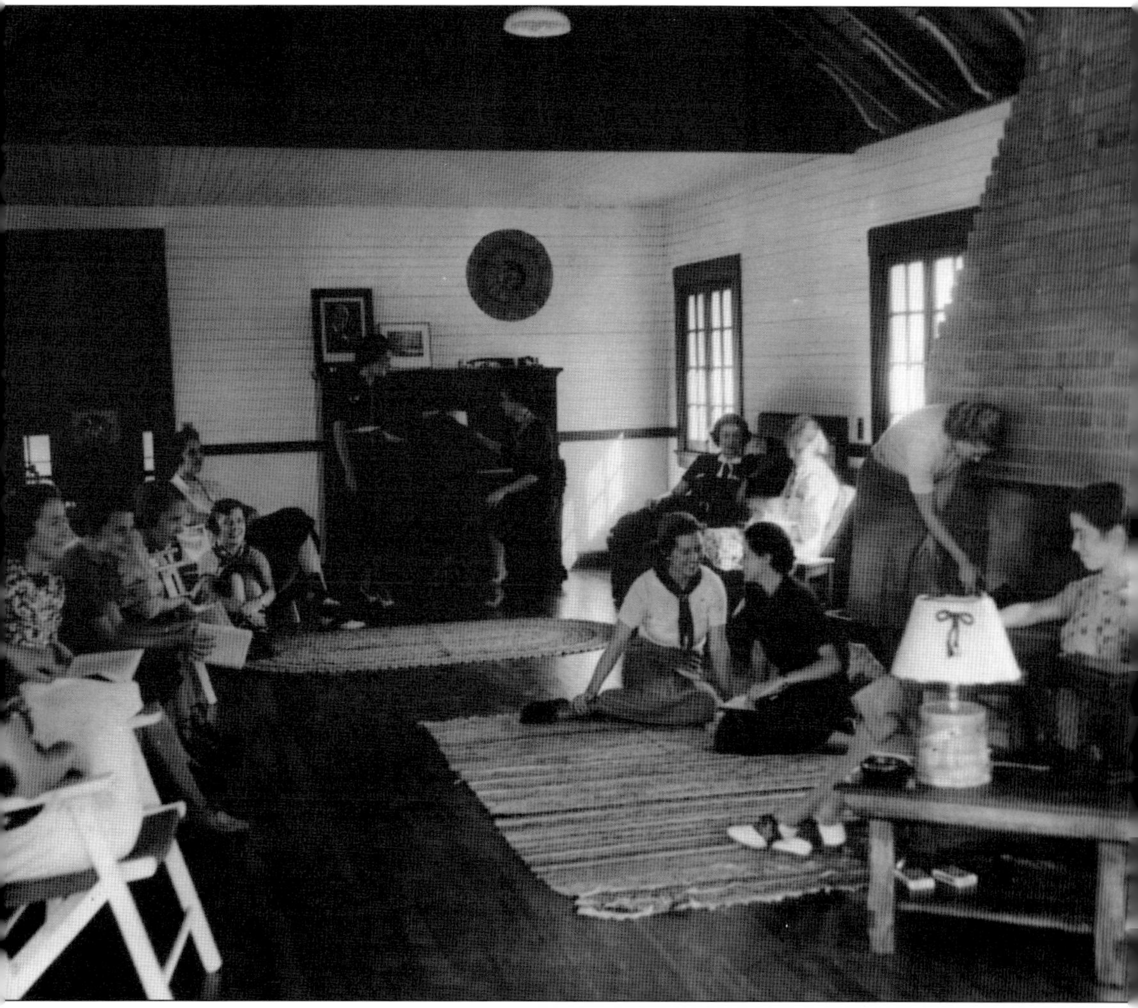

Flora Stone Mather College Students in the Pink Pig. Arguably the most famous building on the farm, the Pink Pig—so named because of its former incarnation as the pig barn—was converted into a cottage-like retreat where social events, dances, and meetings were held. The upstairs loft was converted to a sleeping area that could comfortably sleep a dozen or more students for weekend and overnight gatherings. The wall of the landing that separates the stairs from the loft and the kitchen below bears the names of the original donors to its renovation. The Pink Pig is still a popular destination for Case Western Reserve University student and faculty groups for meetings, retreats, and weekend events.

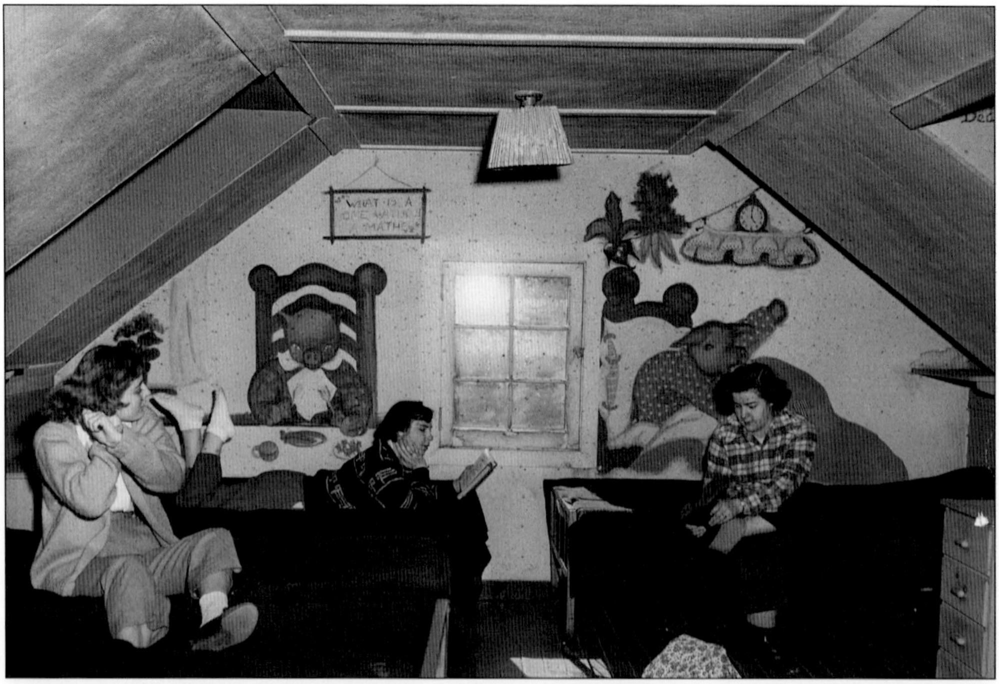

Bedtime in the Pink Pig. These young women are about to retire for the evening in the loft of the Pink Pig. The pig-themed motif is as evident today as it was when this photograph was taken. The sage advice and clever sayings painted on the wall can still be seen.

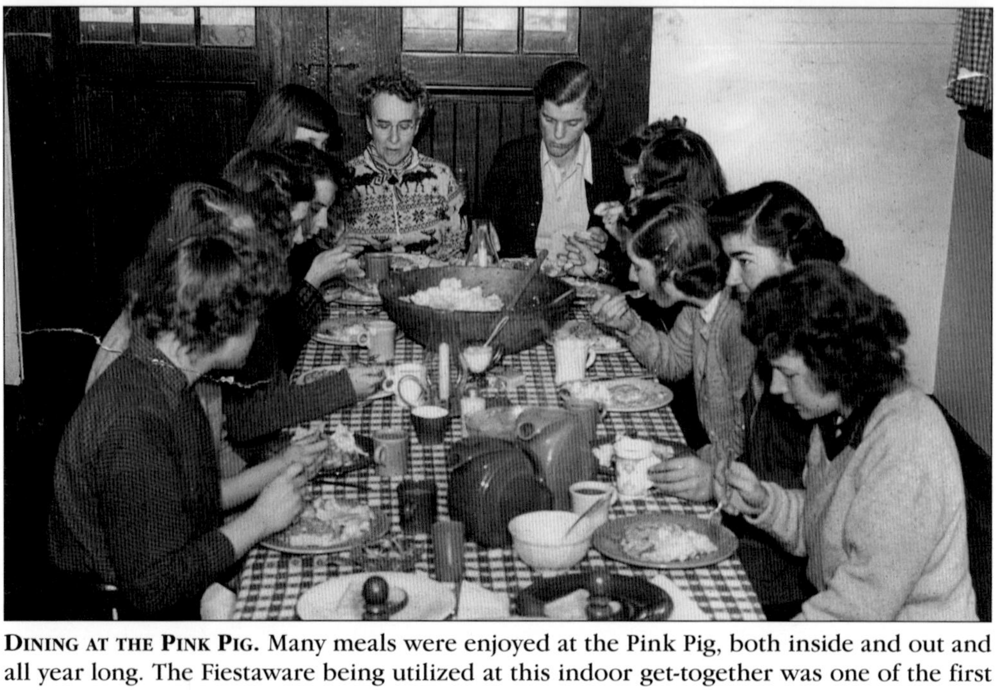

Dining at the Pink Pig. Many meals were enjoyed at the Pink Pig, both inside and out and all year long. The Fiestaware being utilized at this indoor get-together was one of the first donations to the refurbished Pink Pig, and it is enshrined on the donor's list, which is painted on the interior hall wall. Pictured enjoying the meal at the head of the table are beloved dean Elizabeth "Libby" Walker (left) and faculty member Helen Smith.

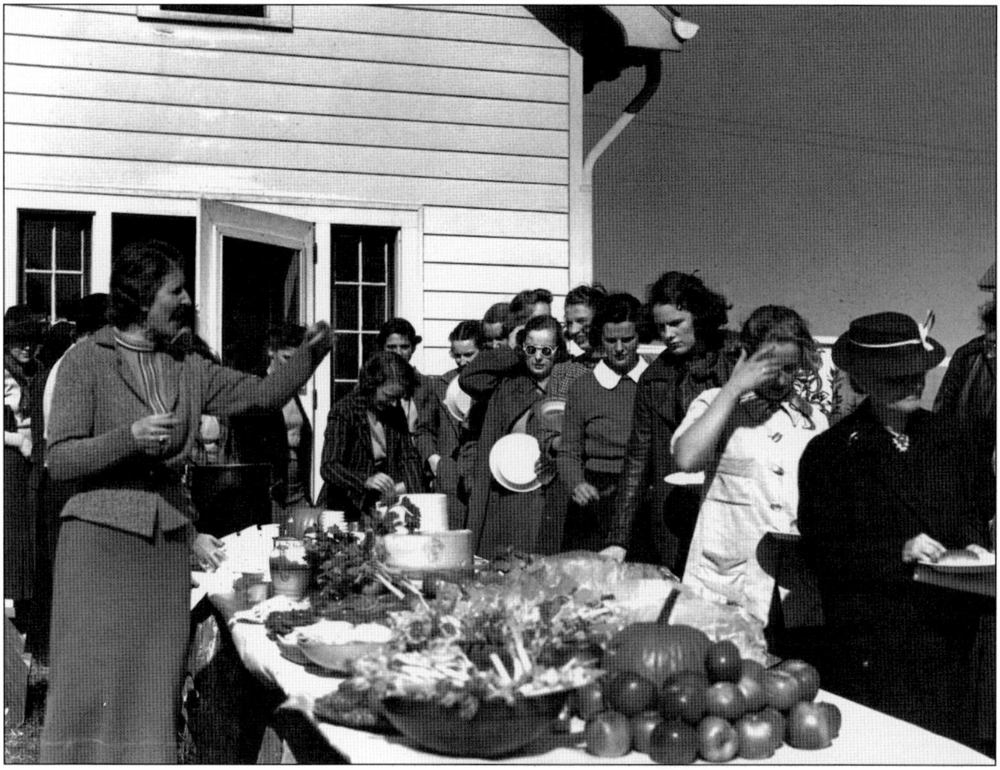

Faculty Farm Days. Outdoor meals were held at harvest time, and the girls would get to enjoy some of the bounty of the farm's produce and their own hard work. Pictured is one of the Faculty Farm Days, when produce from the farm was served. This tradition was reborn in 2010 as the Fall Harvest Festival.

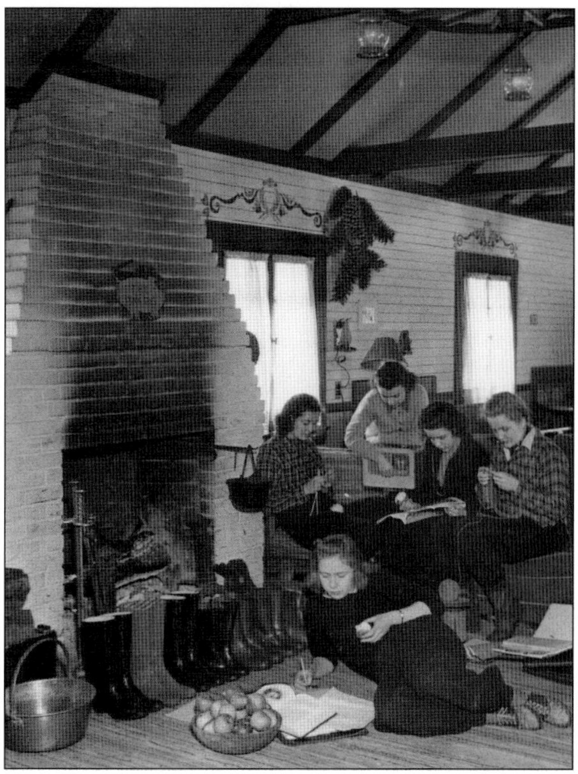

Social Event at the Pink Pig. Young ladies are pictured gathered in the main room of the Pink Pig. Whether the fireplace, the piano, or the company provided the most enjoyment for them is up for debate. Some of the furnishings seen in this picture, as well as the piano, can still be found within the walls of the Pink Pig today.

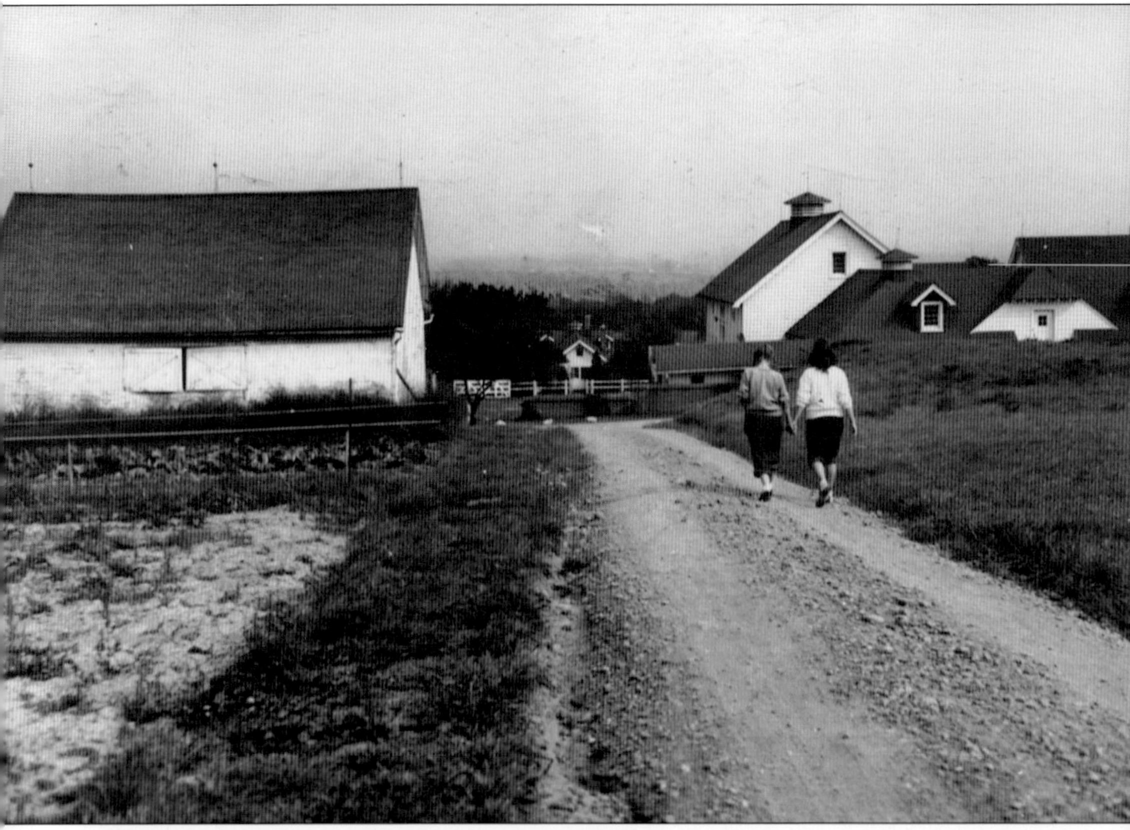

Two Friends Walking. This looks like a perfect view on a perfect day. These two friends are walking hand in hand down the service road that connects the Manor House to the main barn. Near the peak of the property, the magnificent views that led to the name Valleevue can be seen. The white barn on the left still exists and is thought to be the oldest barn on the property (today it is painted red). Friendships, such as the one these girls share, remain a part of what binds Case Western Reserve University students together at University Farm.

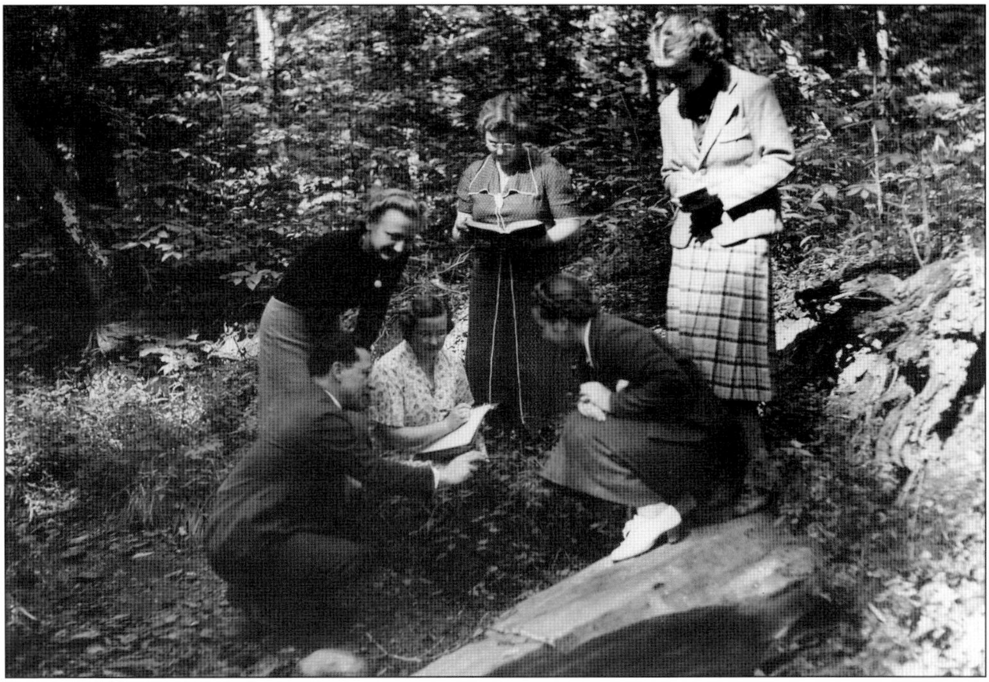

Dr. Bacon with Class. Here, Dr. Franklin Bacon is giving field instruction to a group of young women on the topic of field botany at one of the many unique sites at Valleevue Farm. Note the proper attire worn by all, even when trekking into the woods.

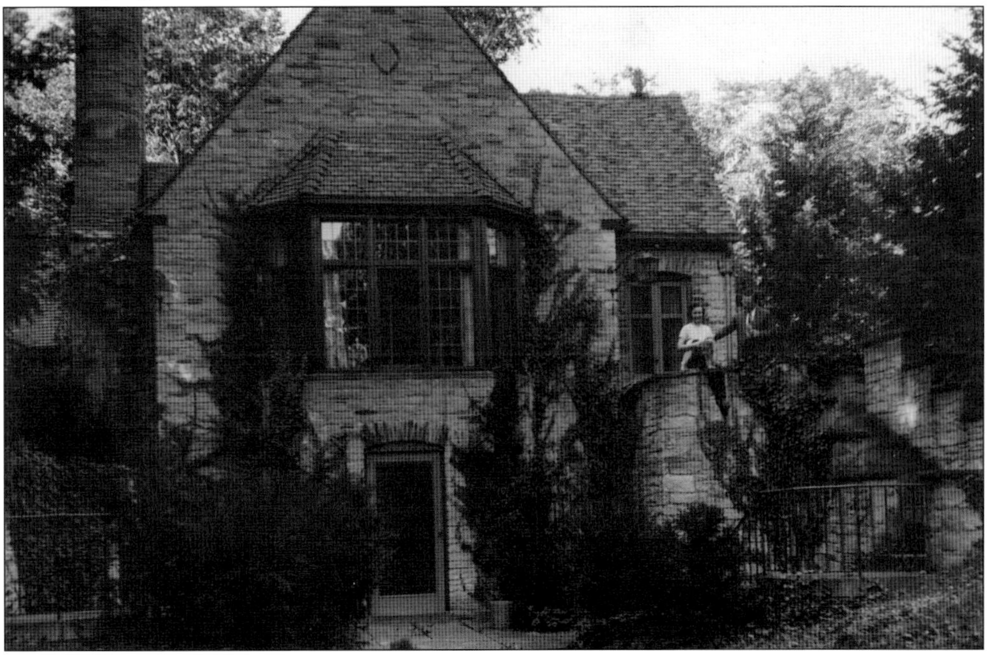

Dr. Bacon at Manor House. Dr. Bacon and his wife with their family dog are shown at their home, the Manor House, during his tenure as farm director. This southern elevation of the house would remain just as pictured for several more decades until it was expanded on in the early 1980s to accommodate one of the university's presidents and his family.

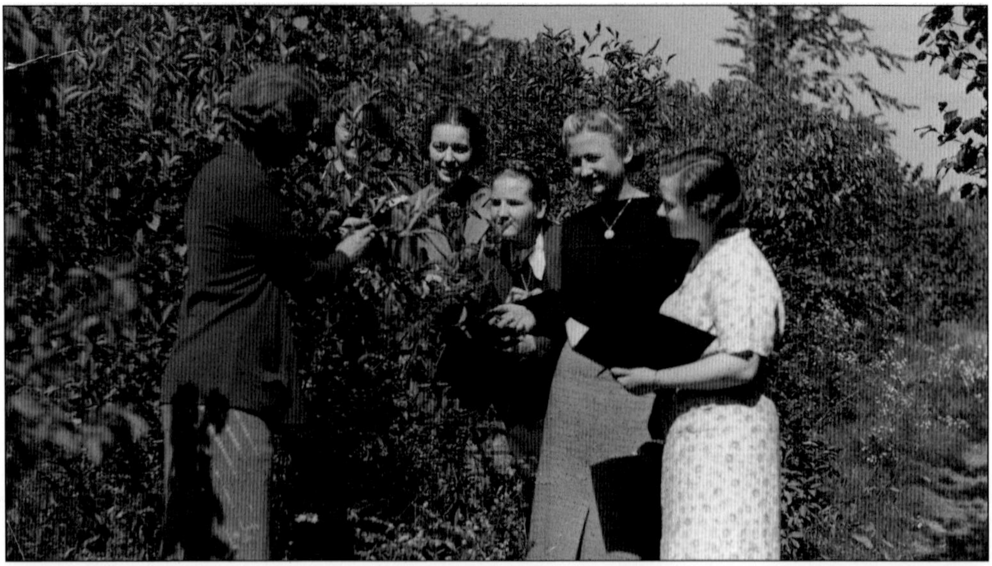

Field Botany. Here is another example of the kind of outdoor education that continues to exemplify University Farm. This class is inspecting the leaves of a fruit tree. Fruit trees once lined the access road that connected the farm property situated on S.O.M. Center Road to the Manor House and Squire Valleevue Farm.

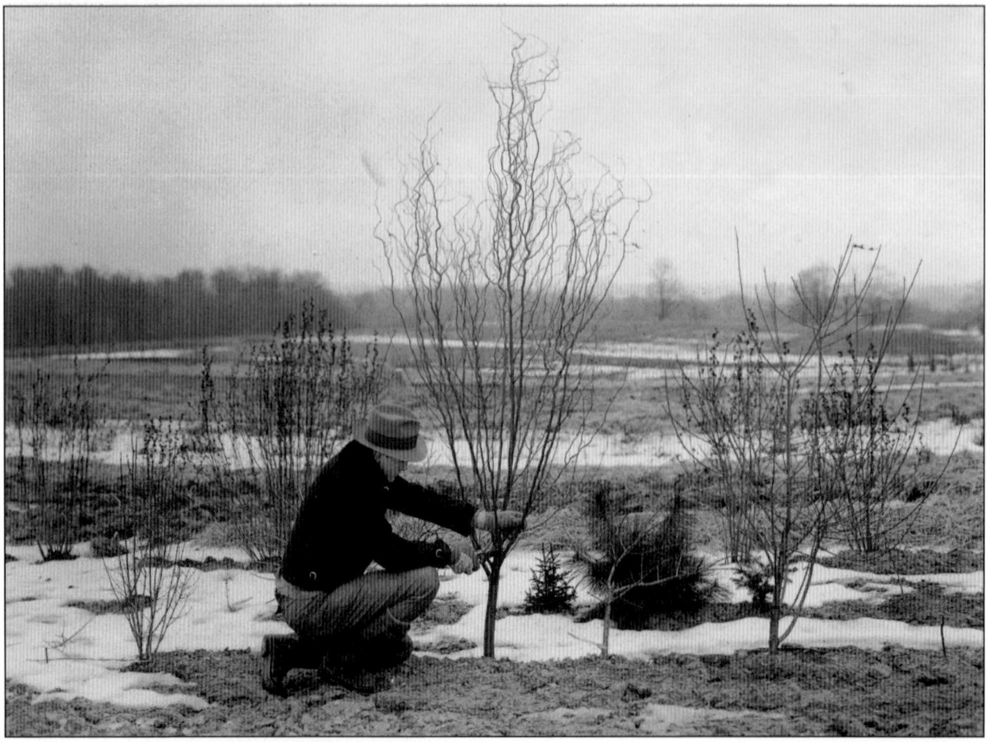

Pruning a Tree. The farm was at one time an impressive repository of cultivated species of both native and nonnative flora. Besides the arboretum and medicinal garden, trees and shrubs of many diverse genera were planted at the farm. Here, a young man prunes a branch off of a tree.

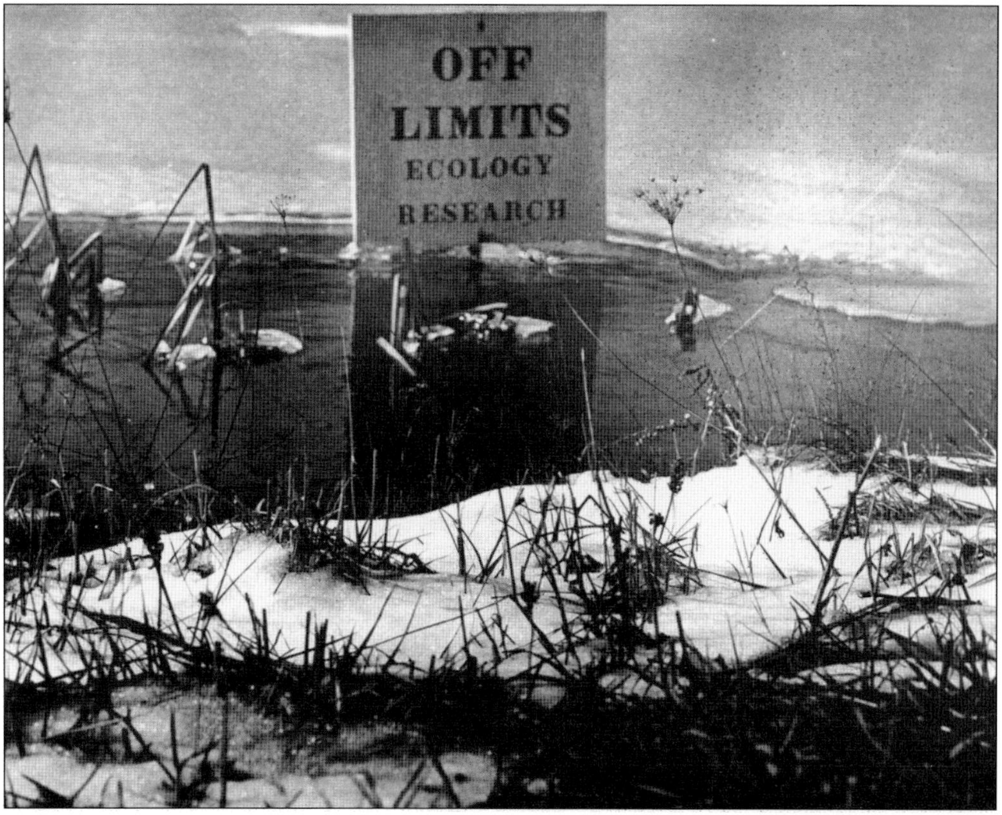

OFF-LIMITS SIGN. Four research ponds were dug at Squire Valleevue Farm between 1956 and 1957: Doc's Pond, Squire Pond, Upper Pond, and Insect Pond. They are still used for research activities, such as studies on the flow of energy and nutrients and the research of fish behavior and plankton distribution. The ponds are also home to large populations of bluegill and largemouth bass. (Andrew Toepfer.)

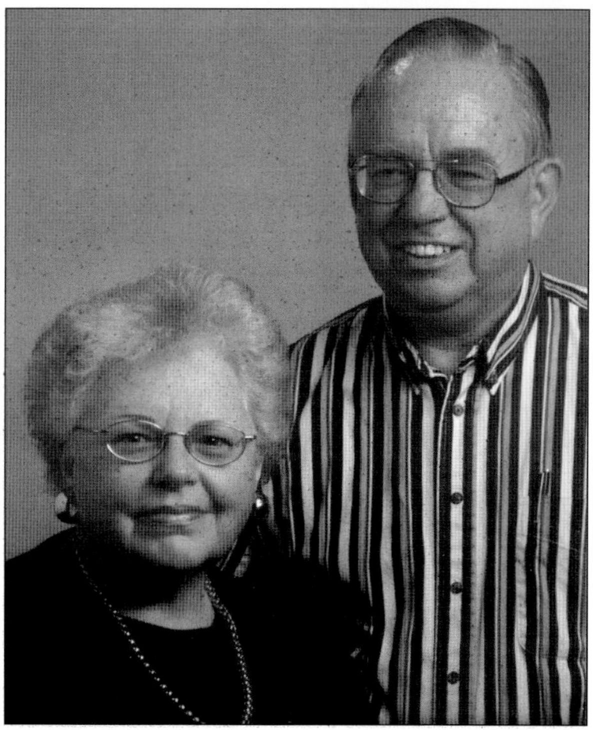

HARRY AND JUNE PALMER. Harry Palmer would serve as the farm's superintendent from 1965 until 2000. June, his wife, would fill the role of secretary during those years. During their tenure, the Palmers would see the size and scope of the farm grow from an area of about 275 acres to one of nearly 400 acres.

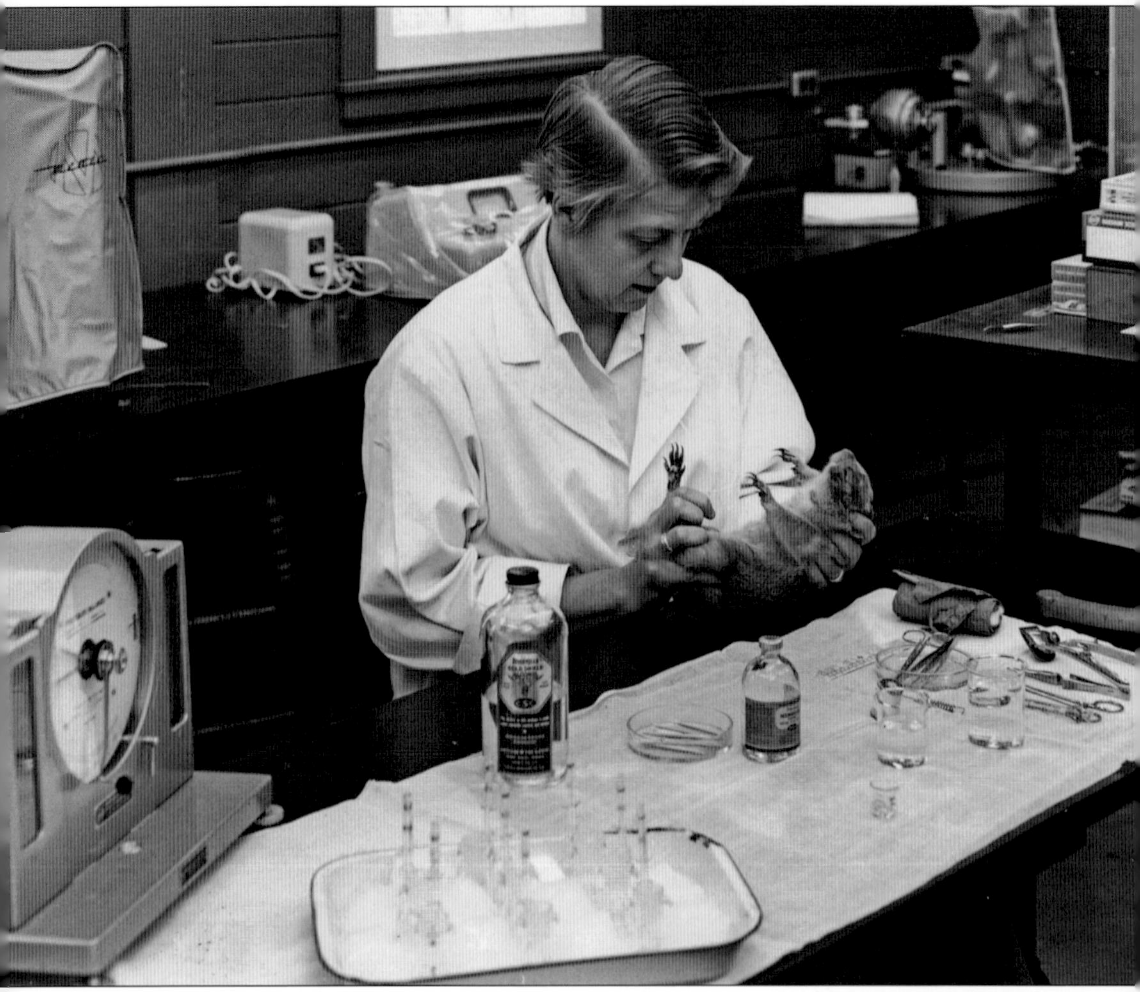

Dr. Foreman in Lab. Dr. Darhl Foreman (now professor emerita and one of the farm's many generous donors) inspects one of the prairie dogs that she researched in her colony at the main barn complex. The lab she is seen working in is now the Foreman Lab, named in honor of her contributions and research done at University Farm. It was originally built with funds obtained by the National Institute of Health (NIH). Dr. Foreman can still be found at her farm office from time to time. Her gifts to the farm have helped to improve both the aesthetics and operations of its facilities.

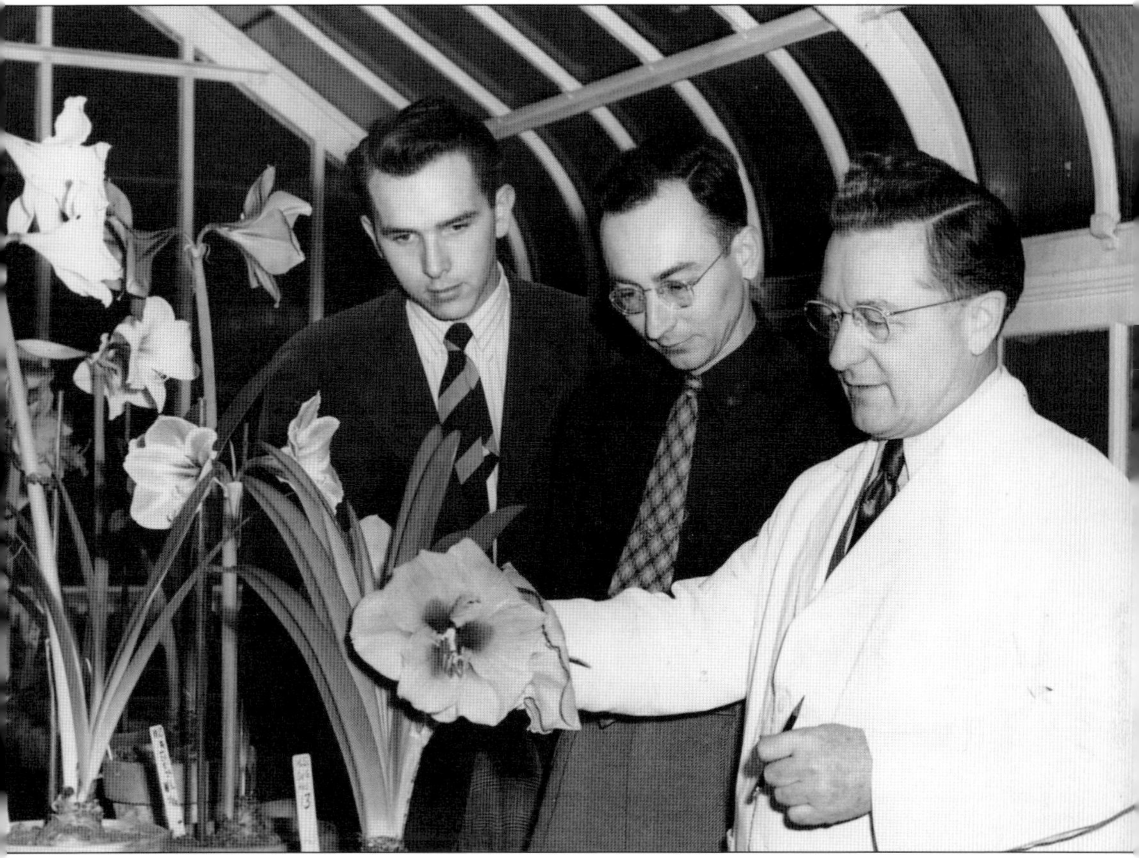

DR. FRANKLIN J. BACON. Dr. Bacon, the farm's first director, is seen here in the research greenhouse examining amaryllis. Dr. Bacon was a friend of the late Andrew Squire. He moved into the Manor House during his term as the farm's director. Research in biology, botany, horticulture, and agriculture was the main focus during his tenure.

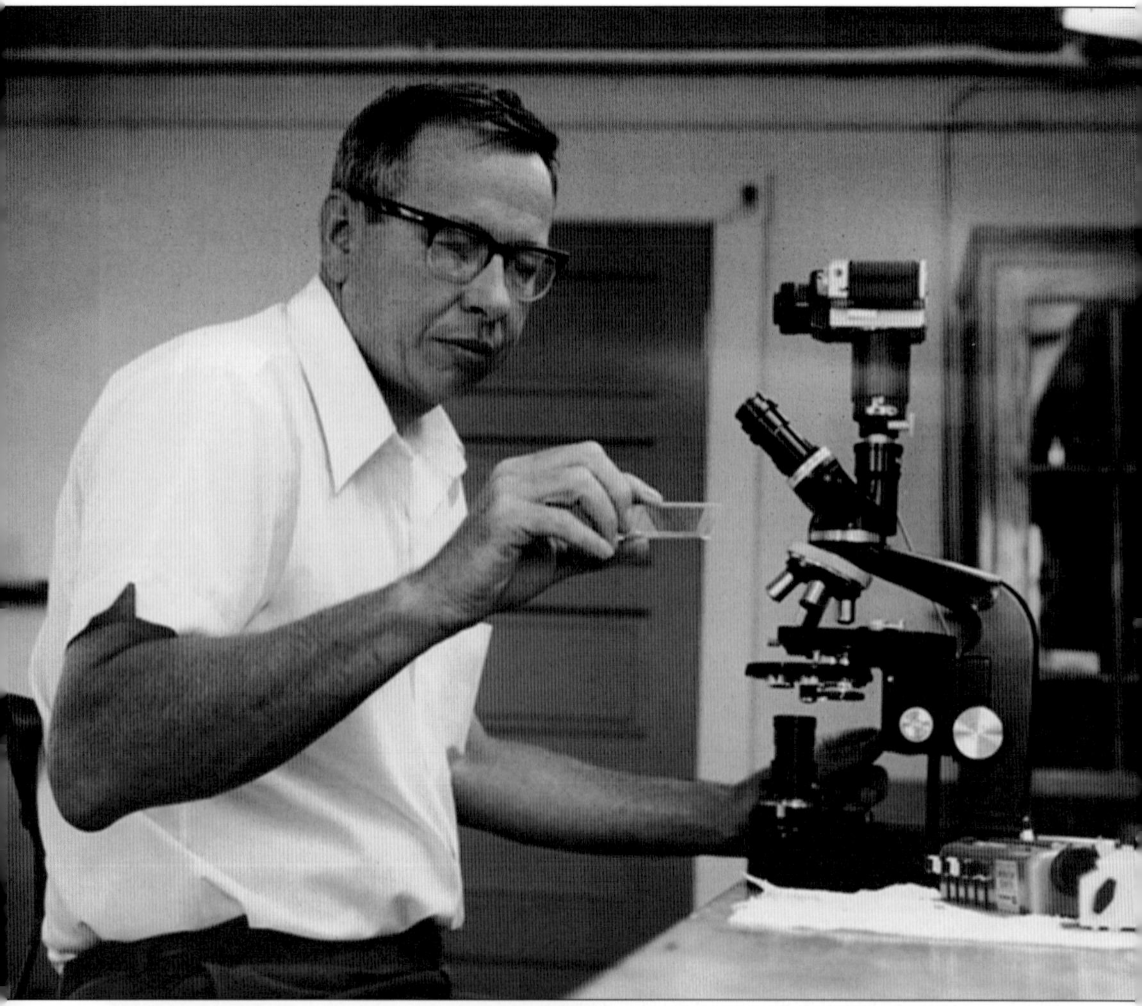

DR. NORMAN ALLDRIDGE IN LAB. Dr. Alldridge was the director of Squire Valleevue Farm from 1960 through 1976, but he continued to do his botanical research, primarily on tomatoes and flax plants, throughout the 1990s. He lived on-site with his family during the years of his directorship at the May Squire House. During his tenure in the director's chair, he would see the farm's usage increase in all aspects: education, research, recreation, and athletics. There is a sign commemorating Dr. Alldridge's service to the students of Case Western Reserve University; it hangs at the entrance to the classroom and labs in the Debra Ann November Research Greenhouse.

Three

University Farm Facilities and Field Station 1973–2009

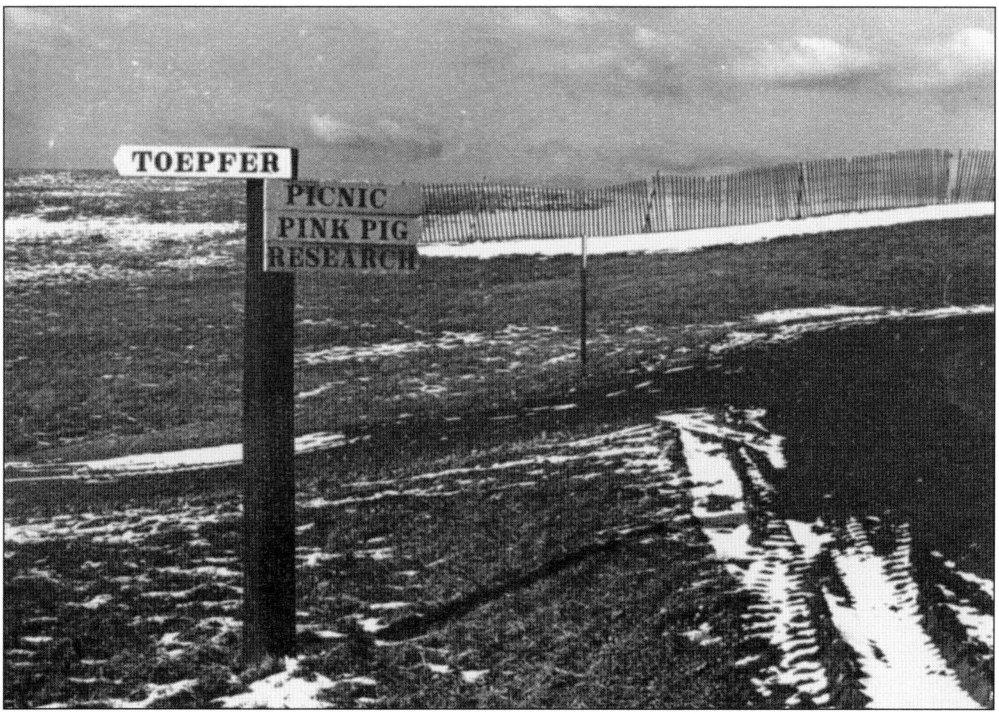

Directional Sign. This early-1970s image depicting a directional sign is a good indication of the diversity of activities and the nature of the multiuse facility that the farm was and still is. This sign points the way to the residence of university president Louis Adelbert Toepfer (the Manor House) and to research and recreational areas, such as the ponds and picnic grounds. (Andrew Toepfer.)

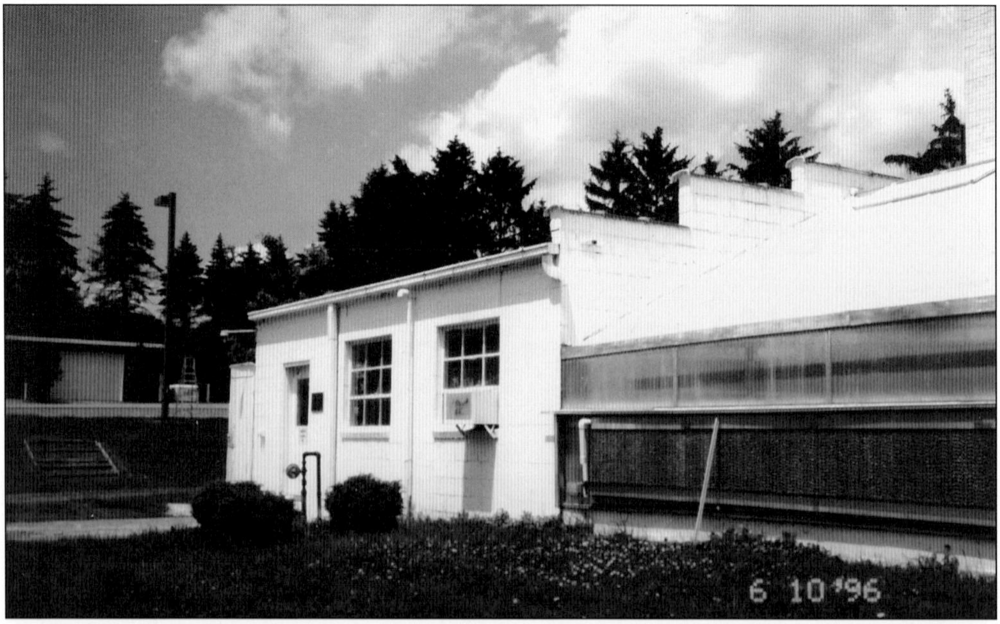

RESEARCH IN GLASS GREENHOUSE. Built in 1947 to aid in the study of botany and horticulture, the greenhouse would be home to research on a variety of plants and insects. Biology department research on tomatoes and genetic studies of flax plants were two of the many endeavors undertaken and that continue to this day. This greenhouse was removed in 2006 to make way for a larger and more energy-efficient greenhouse, the Debra Ann November Research Greenhouse.

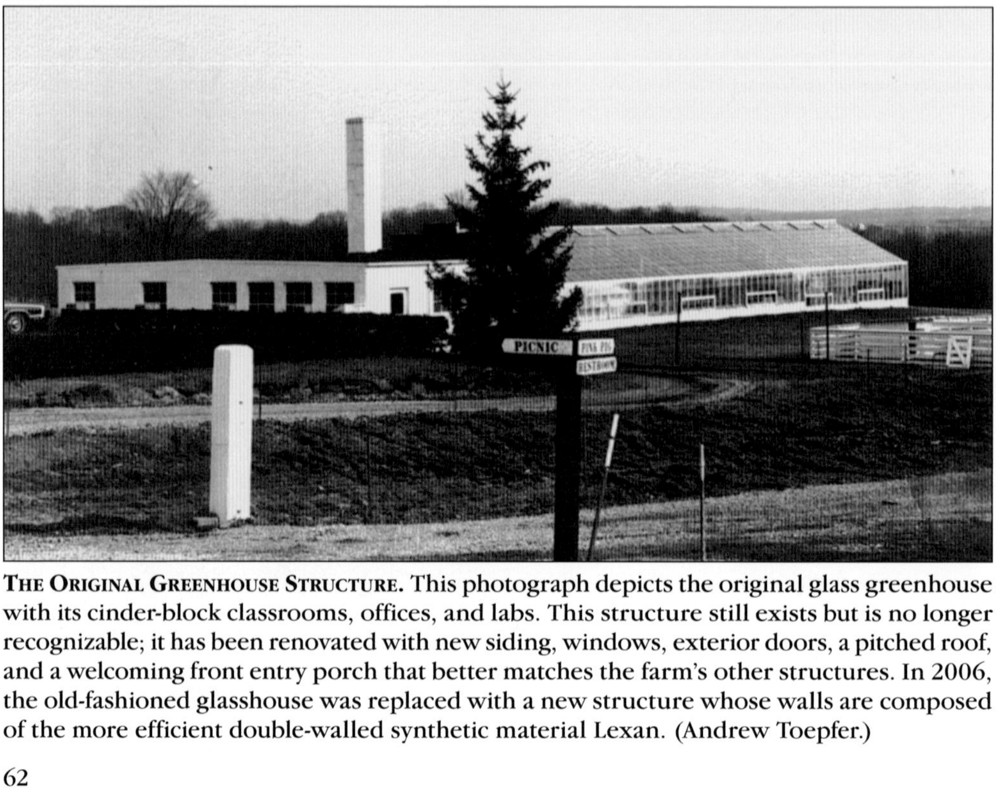

THE ORIGINAL GREENHOUSE STRUCTURE. This photograph depicts the original glass greenhouse with its cinder-block classrooms, offices, and labs. This structure still exists but is no longer recognizable; it has been renovated with new siding, windows, exterior doors, a pitched roof, and a welcoming front entry porch that better matches the farm's other structures. In 2006, the old-fashioned glasshouse was replaced with a new structure whose walls are composed of the more efficient double-walled synthetic material Lexan. (Andrew Toepfer.)

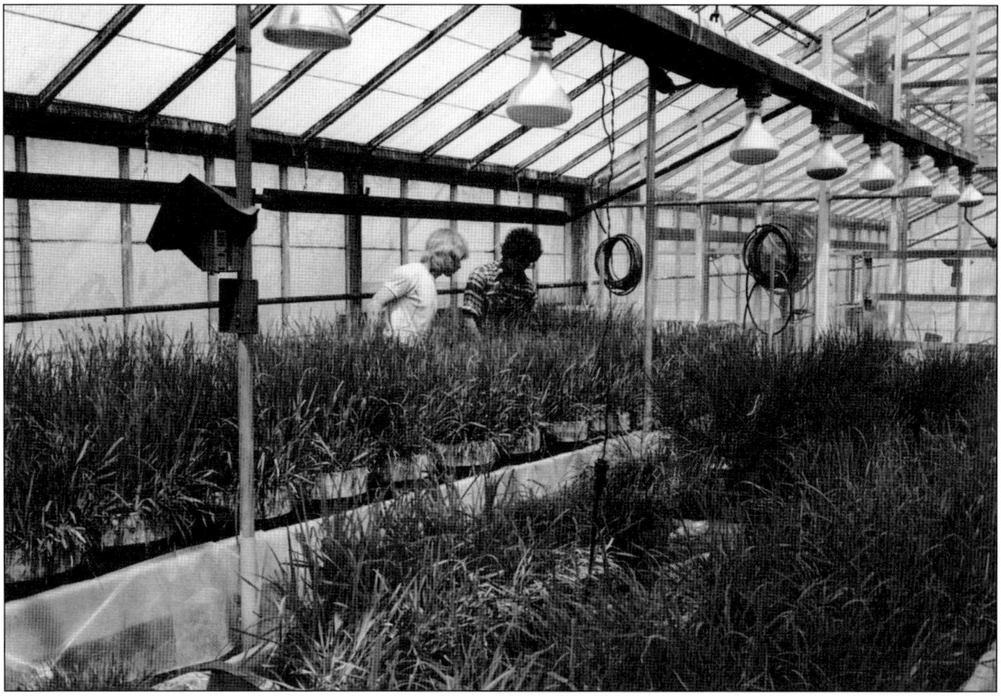

Summer Research in the Greenhouse. Pictured are some students working on a research project in the 1970s. A major obstacle for effective plant growth in the former glass greenhouse was the lack of separation between various research projects. If one group of plants should get infected with a pest or disease, there was no area in which to quarantine them. The current design of the greenhouse allows for such a separation of the plants.

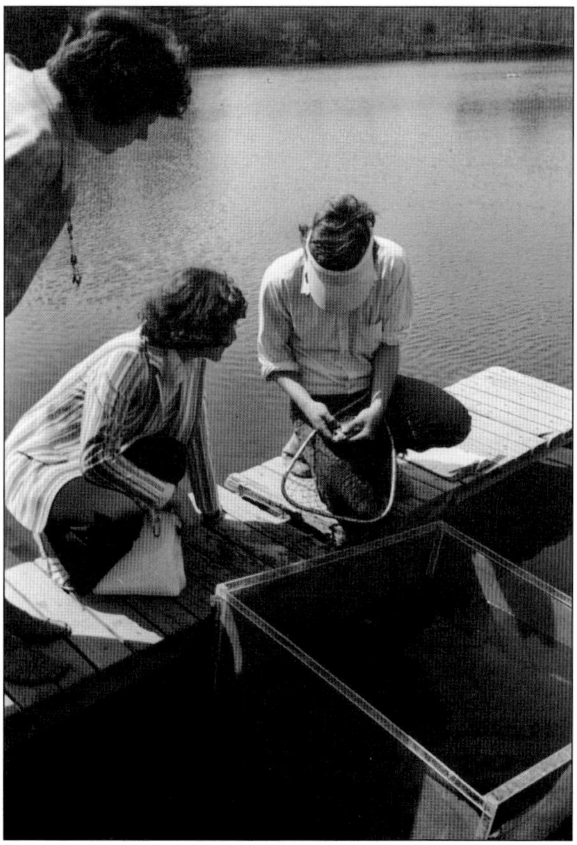

Aquatic Ecology. Here, two students and an instructor, Dr. Sonia Teraguchi (standing on the left), are inspecting pond species at one of the farm's four research ponds. The ponds were installed between 1956 and 1957 by Prof. C.C. Davis and were and still are heavily relied on in the study of aquatic ecology and biology.

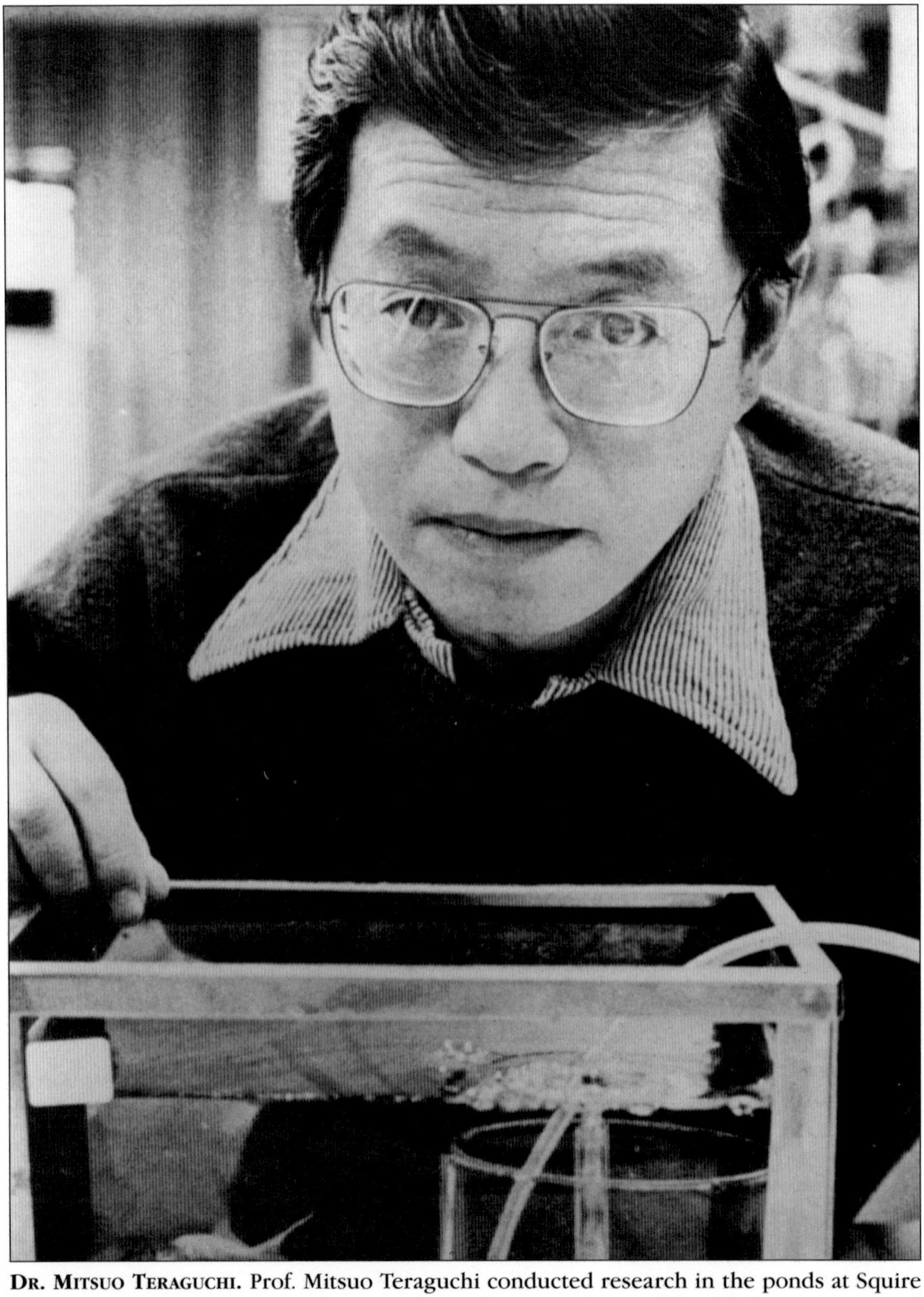

Dr. Mitsuo Teraguchi. Prof. Mitsuo Teraguchi conducted research in the ponds at Squire Valleevue Farm. In 1977, together with his wife, Sonia, he would publish "Structure and Development of Insect Communities in an Ohio Old Field" in the publication *Environmental Entomology*. He would be published again in 1983, with D.E. Stuke and M.M. Noga, in a hydrobiology review.

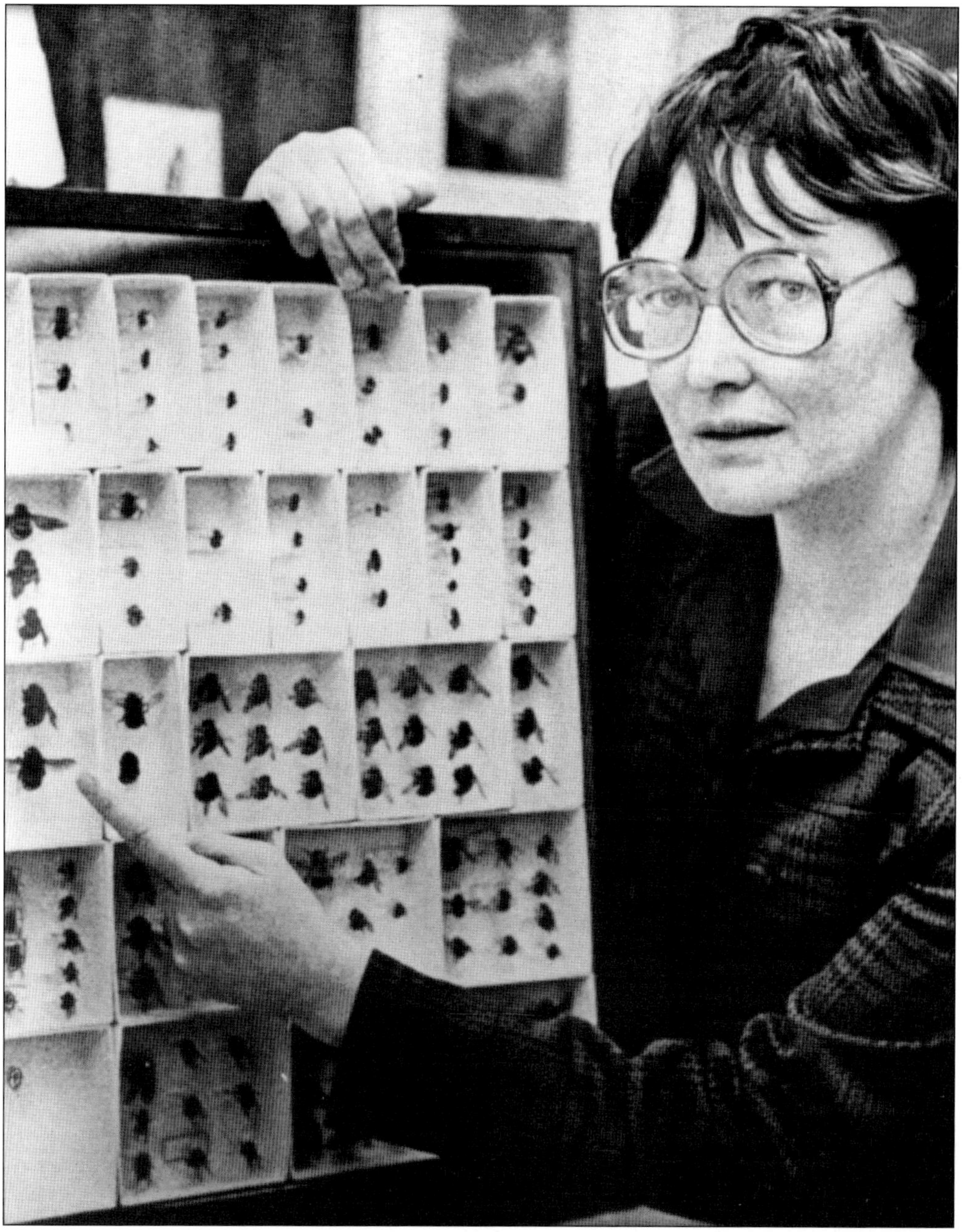

Dr. Sonia Teraguchi. Associate Professor Sonia Teraguchi is seen here with some of her research. In 1975, she would publish "The Meadow Insects of Squire Pond Watershed; Seasonal Changes in Diversity, Abundance, and Trophic Character" as part of the Ecosystem Analysis Program. Her report was submitted to the George Gund Foundation and the Westinghouse Foundation.

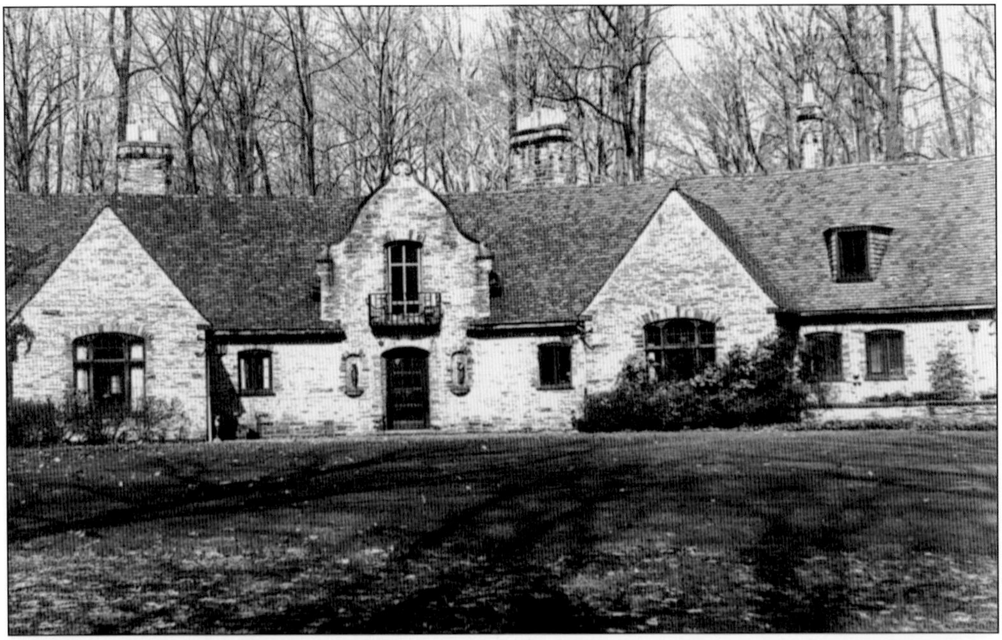

The Manor House as Residence. The Manor House was the residence of university presidents Toepfer during the early 1970s and Ragone in the late 1980s. During this time, the studio apartment addition was constructed and the outdoor staircase was enclosed and became the atrium that connects the Manor House proper with the apartment. Presidents since Ragone live at a house much closer to campus in Cleveland Heights, as the Manor House was deemed too far away from main campus. (Andrew Toepfer.)

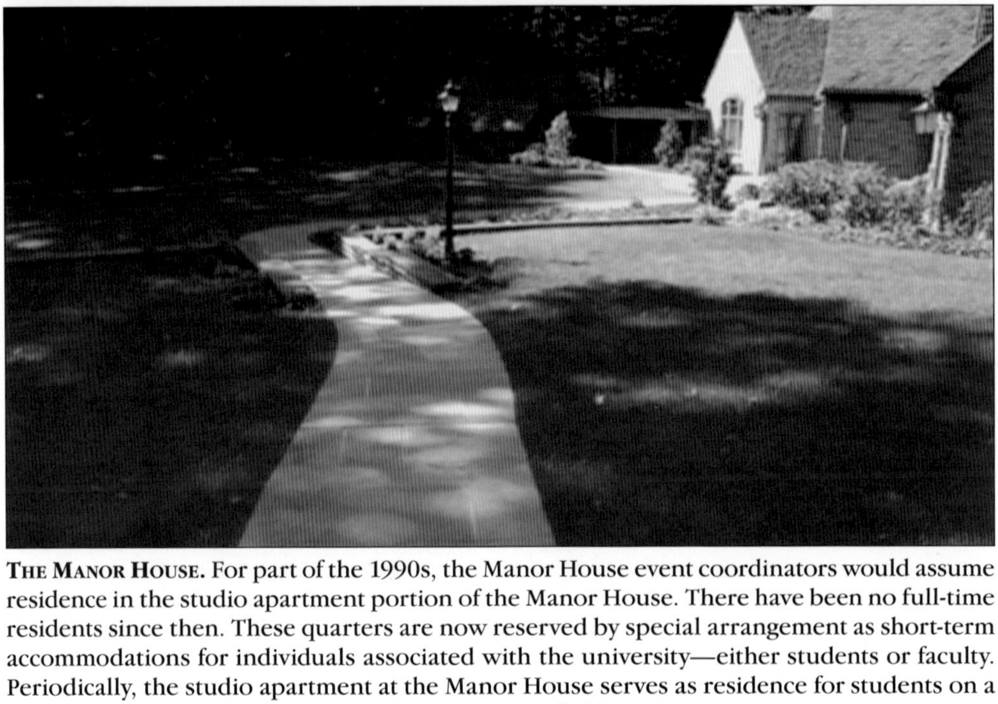

The Manor House. For part of the 1990s, the Manor House event coordinators would assume residence in the studio apartment portion of the Manor House. There have been no full-time residents since then. These quarters are now reserved by special arrangement as short-term accommodations for individuals associated with the university—either students or faculty. Periodically, the studio apartment at the Manor House serves as residence for students on a short-term basis while they are conducting on-site field studies.

Four

WADE'S VALLEY RIDGE FARM 1905–1977

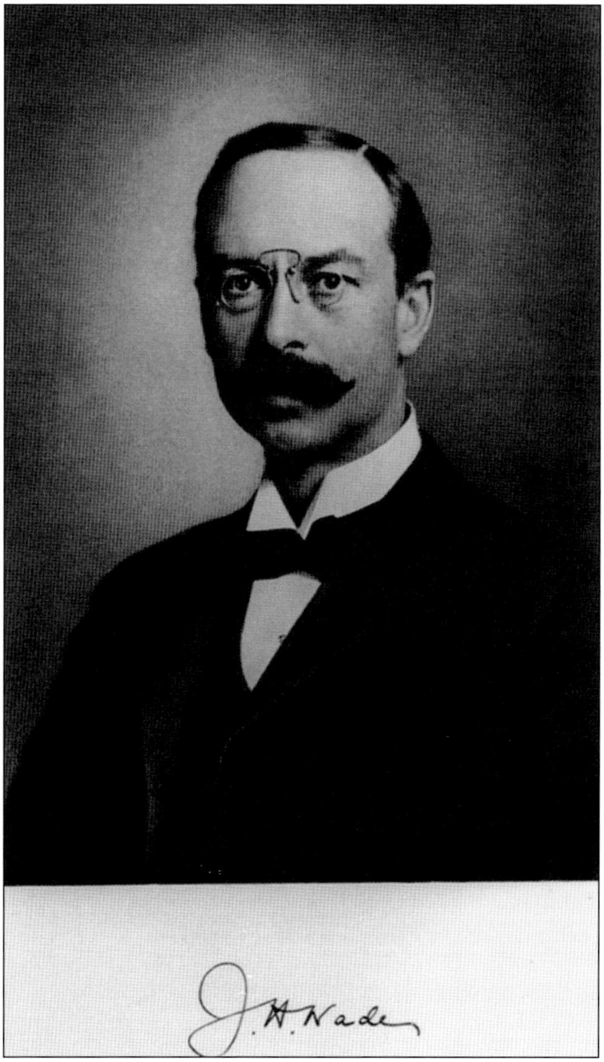

JEPTHA HOMER WADE II. Businessman, financier, philanthropist, and gentleman farmer J.H Wade—or Homer, as he was known to his friends—was the heir to the Western Union Telegraph fortune, which was the legacy of his grandfather, Jeptha Homer Wade. J.H. Wade II was an extremely successful businessman in his own right, serving as executive of more than 45 financial institutions and businesses over the course of his life. (WRHS.)

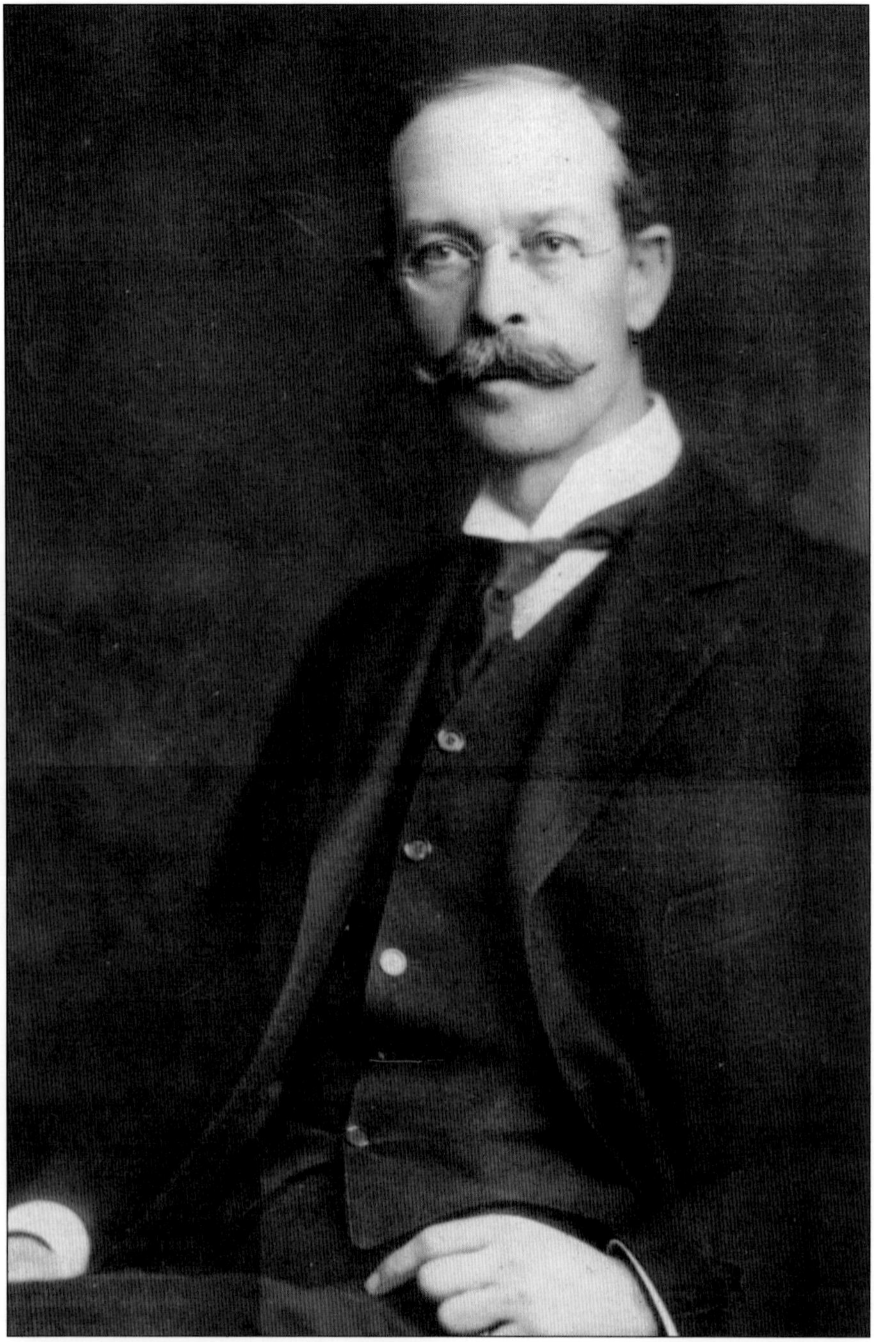

Jeptha Homer Wade II Later in Life. By the end of his nearly 70-year-long life, J.H. Wade II would donate millions of dollars to multiple Cleveland-area institutions. He would contribute both time and resources to organizations such as the Cleveland Art School, Lakeside Hospital, the Protestant Orphan Asylum, Western Reserve University, and the Western Reserve Historical Society, which is currently the repository of the Wade family papers. He would also serve as the Cleveland Museum of Art's first vice president and, in 1920, as president of the board of trustees until 1926. (WRHS.)

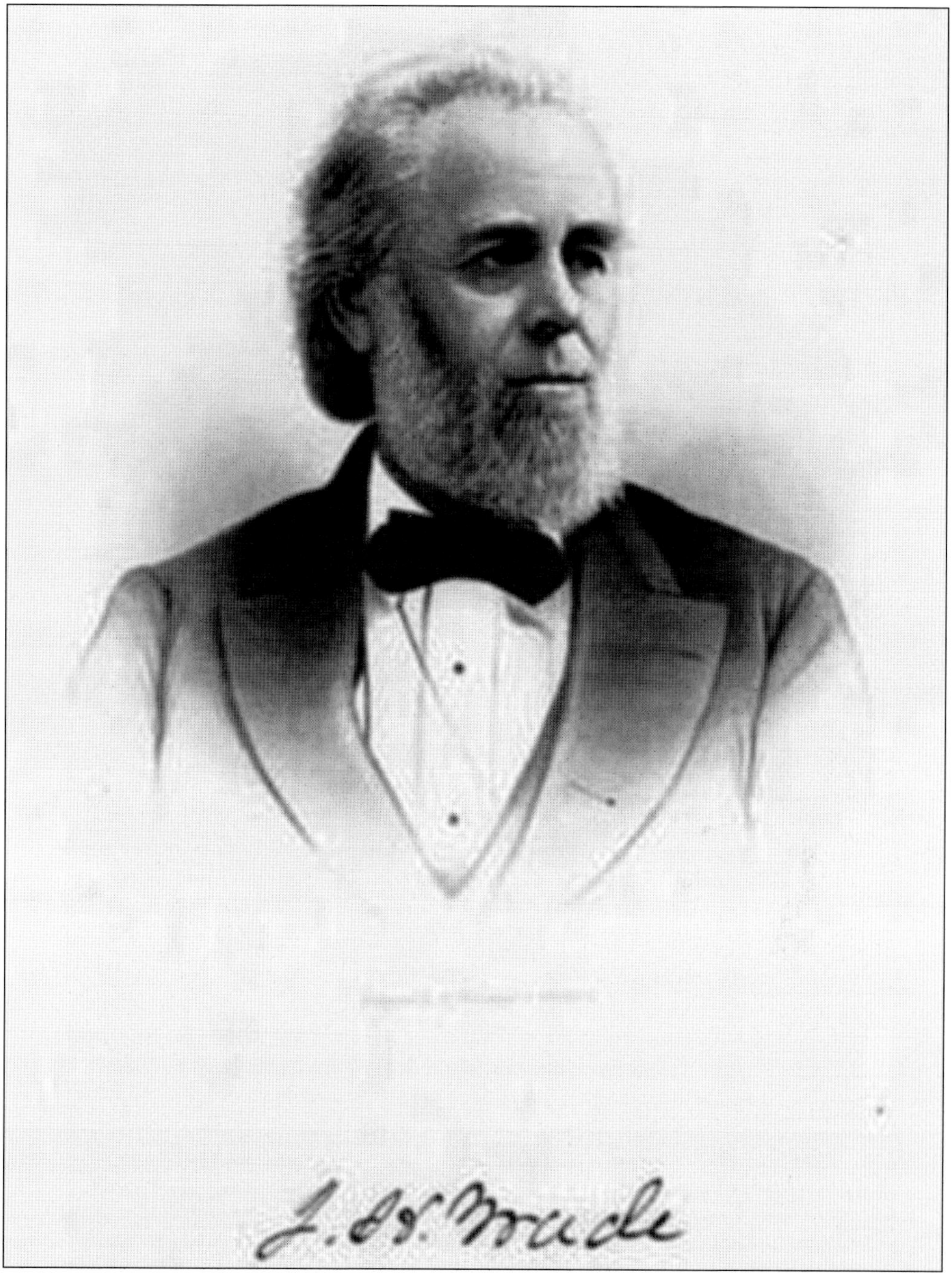

Jeptha Homer Wade. Grandfather of Jeptha Homer Wade II, he would make his fortune by consolidating a number of telecommunications companies into the telegraph giant Western Union. Jeptha Wade was born in 1811 in Romulus, New York, and was an artist before developing an interest in communications. He would be a great influence on his grandson, including passing on his practice of philanthropy. J.H. Wade II had the Wade Chapel, complete with a grand window designed by Louis C. Tiffany, in Lakeview Cemetery built in his grandfather's honor and memory.

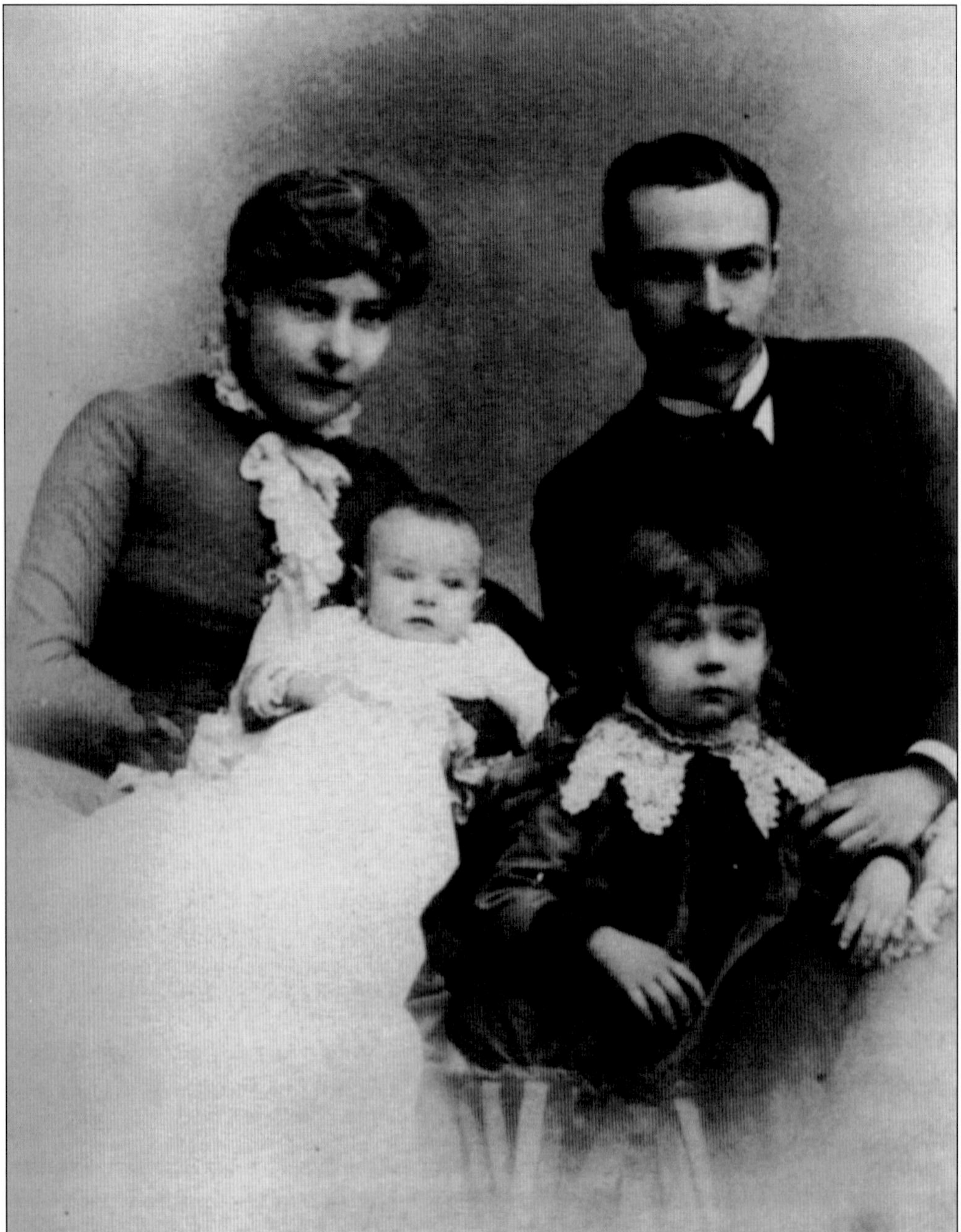

The Wade Family. The Wade family, under the guidance of J.H. Wade II, would become one of Greater Cleveland's most generous benefactors. Throughout his life, he would support dozens of worthy institutions and charities. He donated the land to build the Cleveland Museum of Art, and his personal collection of jewels and gems can still be seen in the Cleveland Museum of Natural History. Many Cleveland-area landmarks, including Wade Chapel, Wade Lagoon, and Wade Oval, still honor the legacy of philanthropy this man and his family left behind. Pictured is J.H. Wade II as a child with his father Randall, mother Ana (McGaw), and sister Alice. (WRHS.)

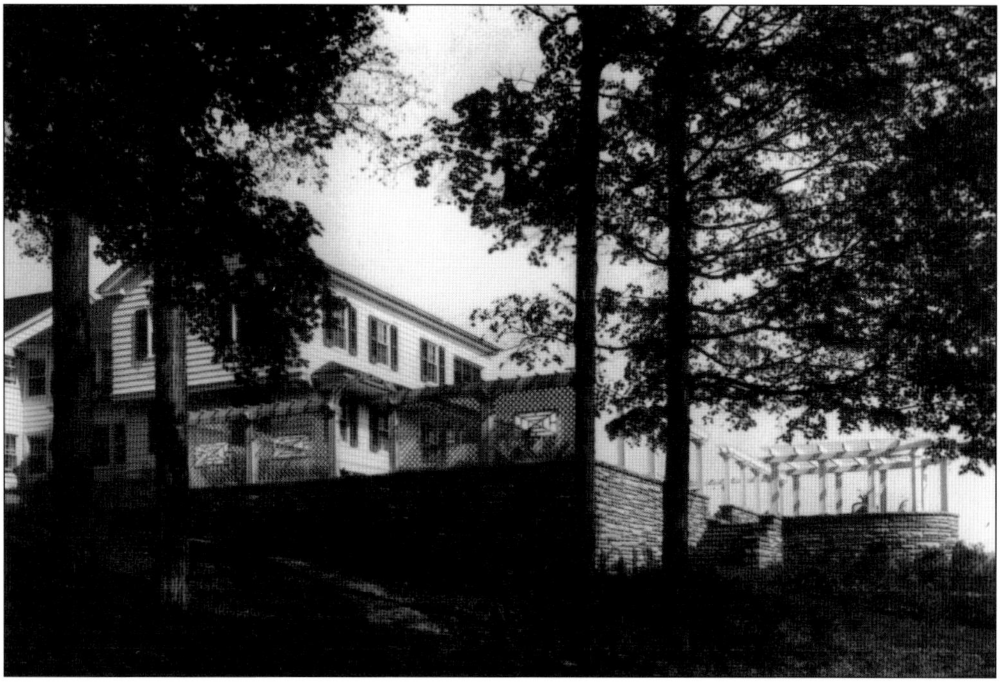

J.H. Wade II's Impressive Estate. The architectural firm of Hubbell and Benes designed this neo-Colonial mansion overlooking the banks of nearby Chagrin River. Just as Wade would share a decades-long collaboration with landscape architect Warren Manning, he shared a lengthy relationship with Hubbell and Benes. The architects would go on to design the building for the Cleveland Museum of Art, with which J.H. Wade II was heavily involved. The Wade estate at Valley Ridge Farm was said to have been filled with Chippendale, Sheraton, and Empire furniture. Some of the Wades' extensive art and antiquity collection also was housed here. Ellen Wade chose white, pink, cream, and green as her decorating scheme. (Both, WRHS.)

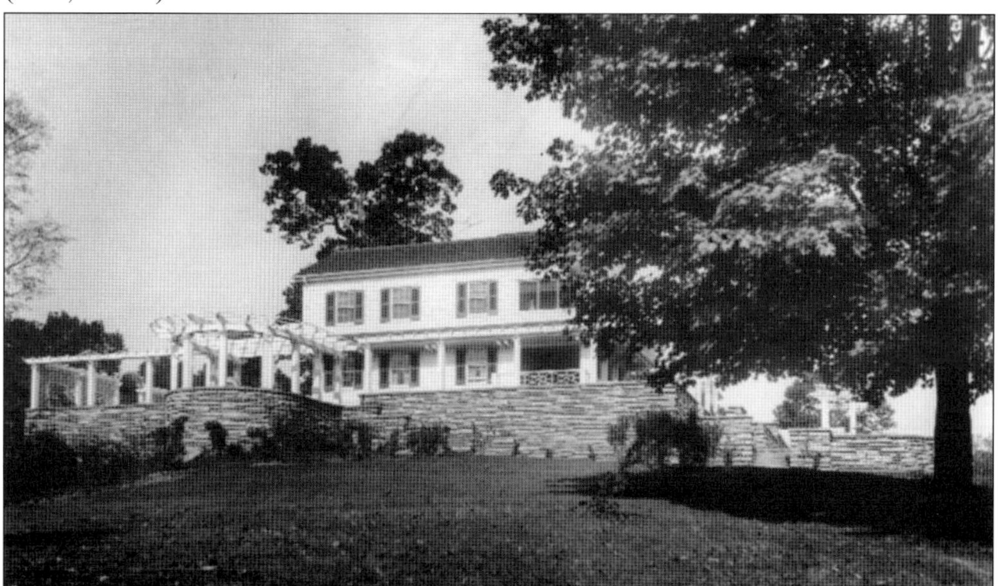

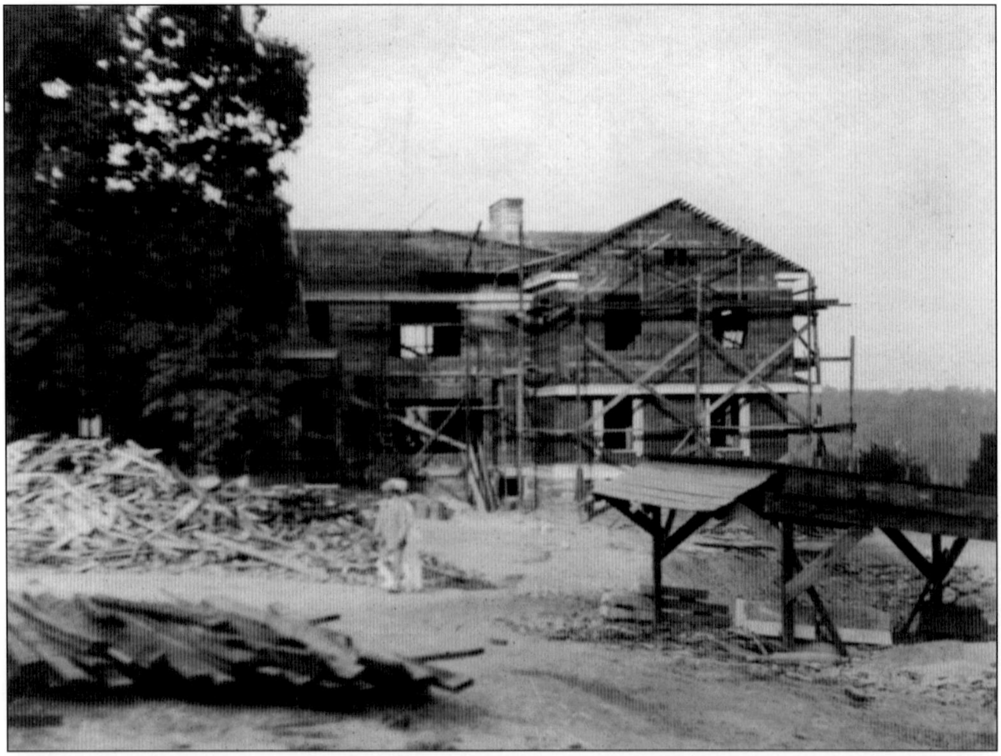

HOUSE UNDER CONSTRUCTION. These photographs from the Wade family's personal collection are dated May 26, 1909, and picture the mansion during its construction phase. The entire compound would be fully modern for its time. All of the buildings, including the dairy, horse, pig, and chicken barns had full electric service and plumbing. This would be another home that J.H. Wade II would have commissioned by Hubbell and Benes; he hired them to design and build his Thomasville, Georgia, estate (Mill Pond) as well. Tragically, the mansion at Valley Ridge Farm would only stand for 20 years before being destroyed by fire in 1929. (Both, Elizabeth Wade-Sedgwick.)

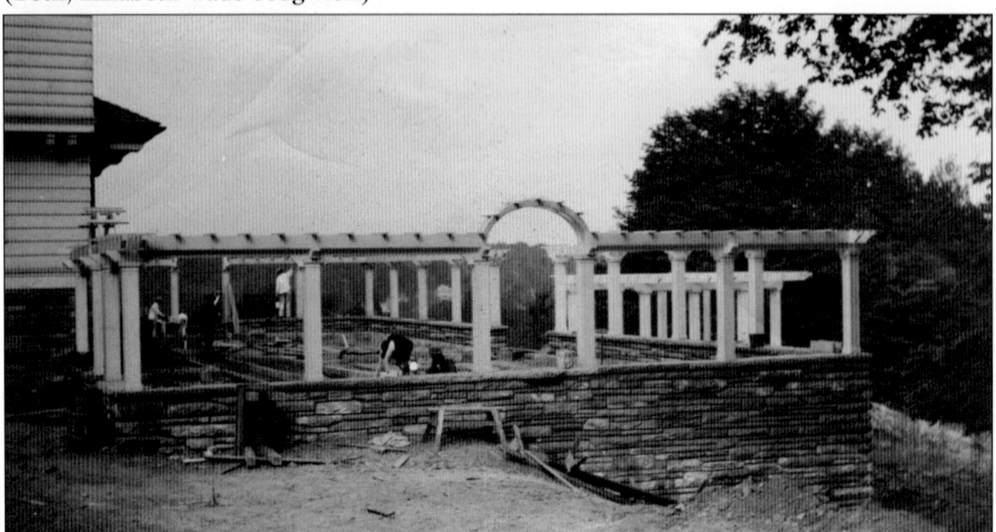

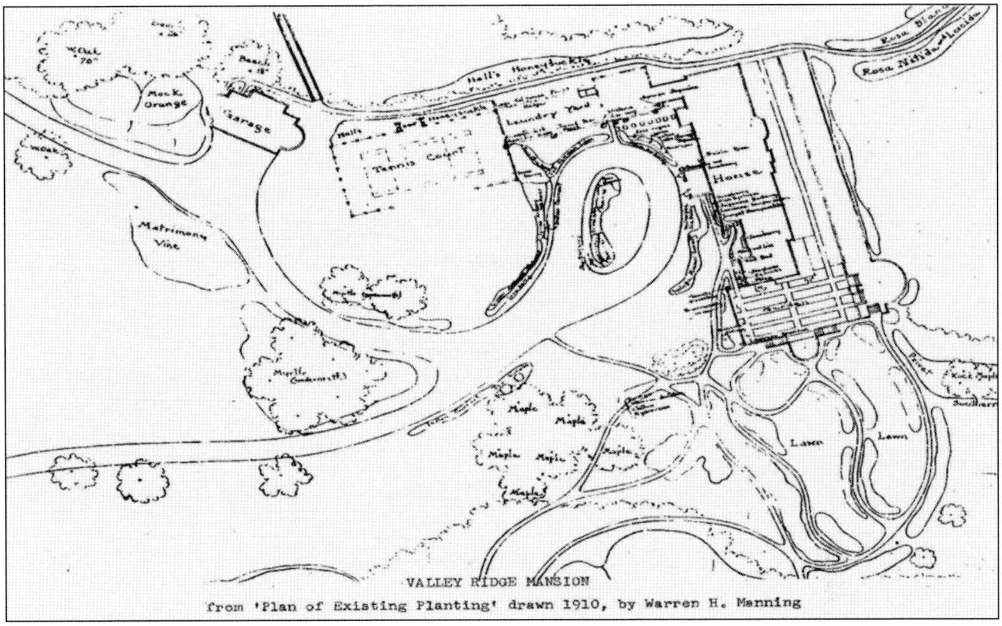

MANNING DRAWING OF HOUSE. The Wade family mansion, with its elaborate features and gardens, can be seen in this portion of a 1910 landscape drawing by Warren H. Manning. This is a detail of the larger plan on pages 66 and 67. To this day, some of the progeny of the plants depicted in this drawing thrive in the general vicinity of where the house once stood. Hydrangea, barberry, and spirea can be easily found. (UIA.)

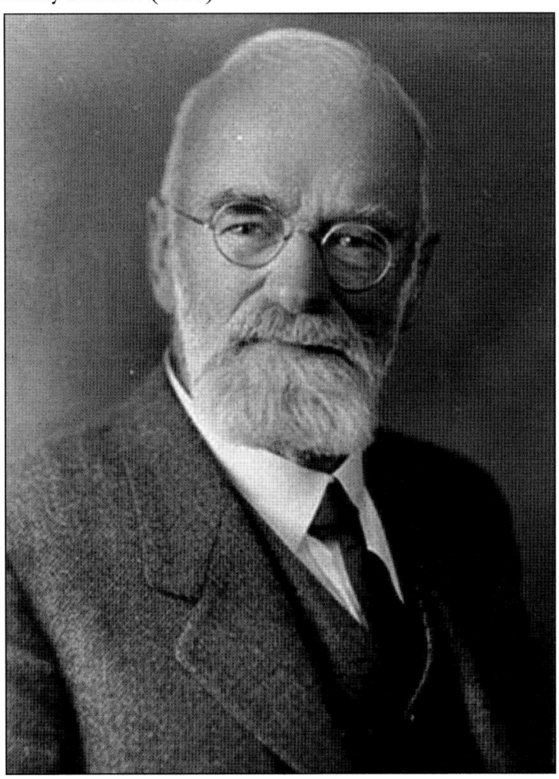

AMERICAN LANDSCAPE ARCHITECT WARREN H. MANNING. Manning would form a working relationship with J.H. Wade II that lasted over 20 years, and he created one of his masterpieces at the valley estate. Manning created hidden gardens and paths, cut vistas through the woods to allow for a view of the river below, and orchestrated all of the diverse topography into one cohesive work of art.

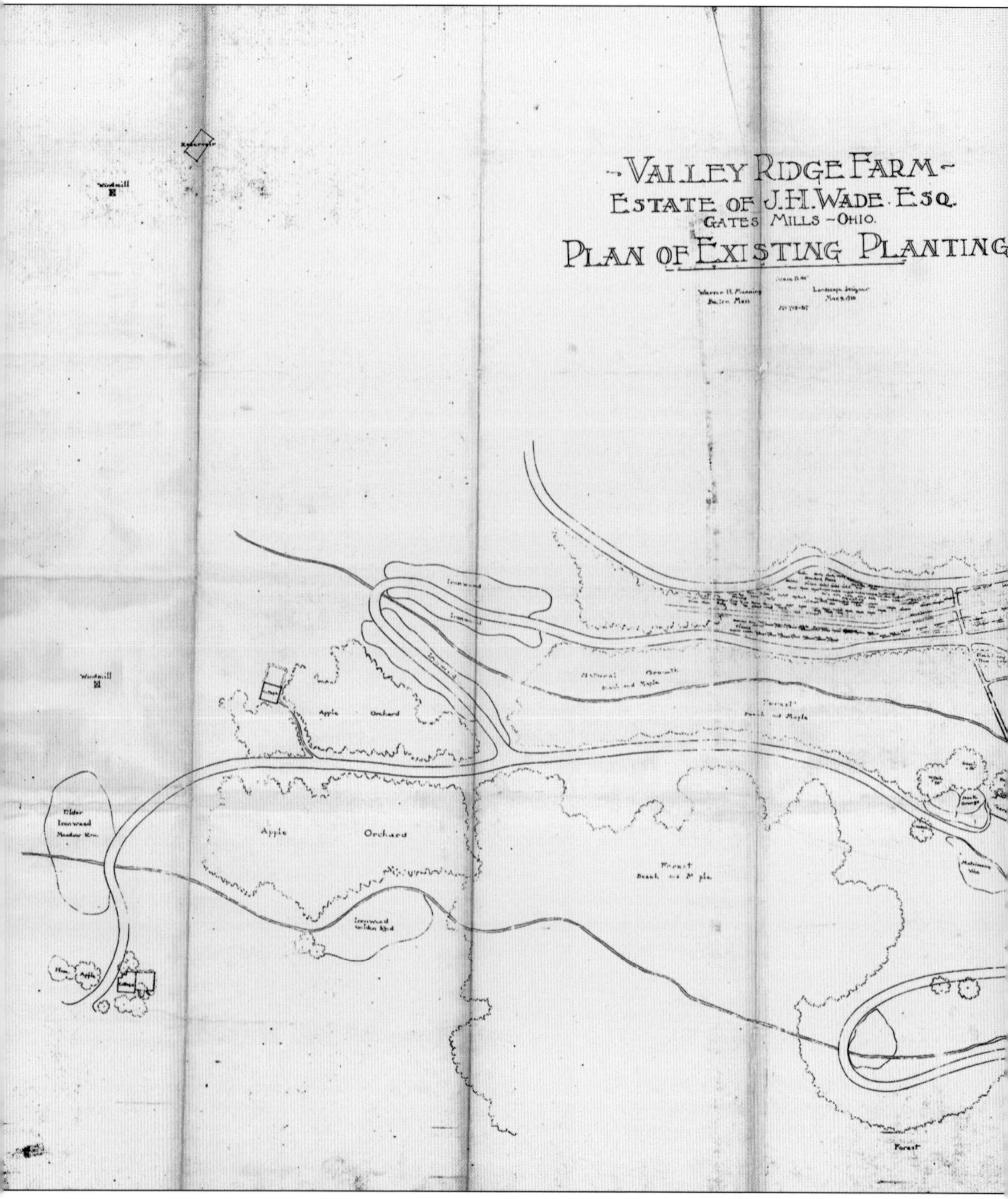

LANDSCAPE PLAN FROM 1920 BY WARREN H. MANNING. Over 100 of these drawings were done by Manning for Wade's Valley Ridge Farm between 1905 and 1925. This drawing highlights some of the man-made features and gardens, such as the Labyrinth, Chestnut Point, and the Garden of Eden, which contained hundreds of species of fruit-bearing trees, shrubs, and vines.

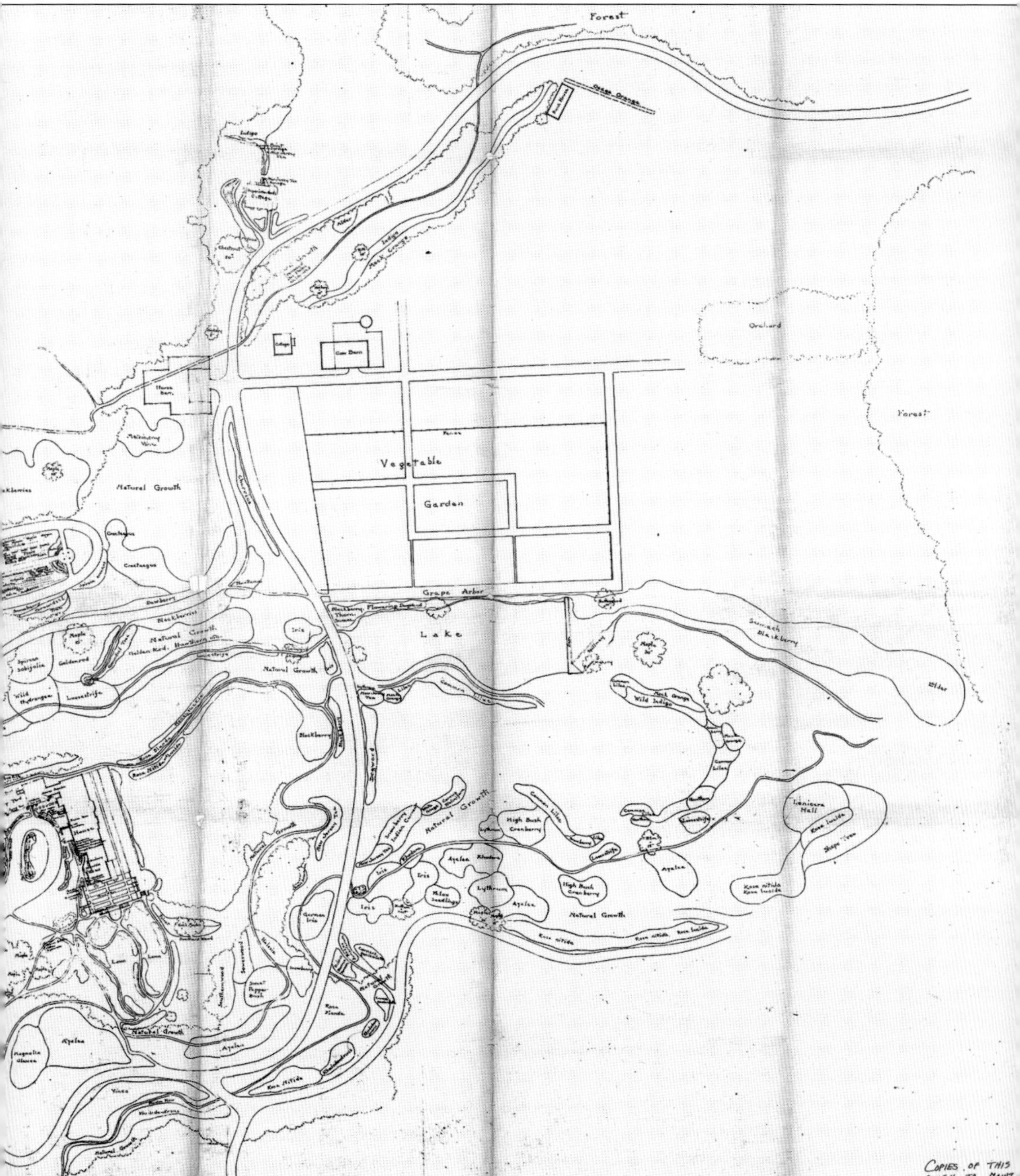

Most of the features of this drawing no longer exist, however, large swaths of *Aegopodium podagraria* (snow-on-the-mountain) and *Vinca minor* (periwinkle) carpet portions of the forest floor, remnants of the former landscaped splendor. Efforts are currently being made and plans are underway to restore some aspects of this historical landscape. (UIA.)

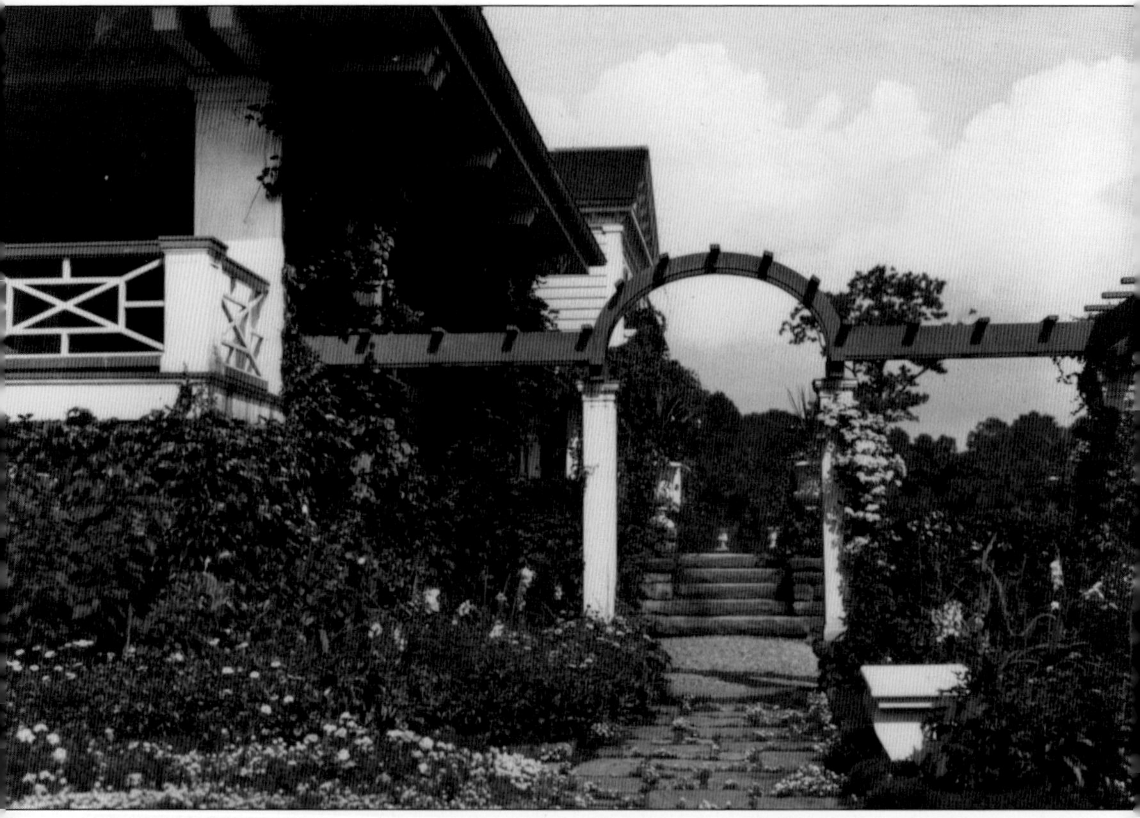

Valley Ridge Farm Gardens. This c. 1915 photograph includes the gardens and ornate latticework that graced the immediate surrounds of the house. Warren Manning would design elaborate gardens making use of many native plant species, and he allowed room for his choices to grow into their natural forms. He did not subscribe to the Victorian era's approach to landscape design, which relied heavily on unnatural geometric shapes and was still quite prevalent at this time. Manning was also quite progressive, frequently using female designers as his subcontractors. Some frequent collaborators on Manning's designs were noted garden designer Ellen Biddle Shipman and Violet Ethelwyn Harrison (who would later serve as superintendent at Squire Valleevue Farm). (Elizabeth Wade-Sedgwick.)

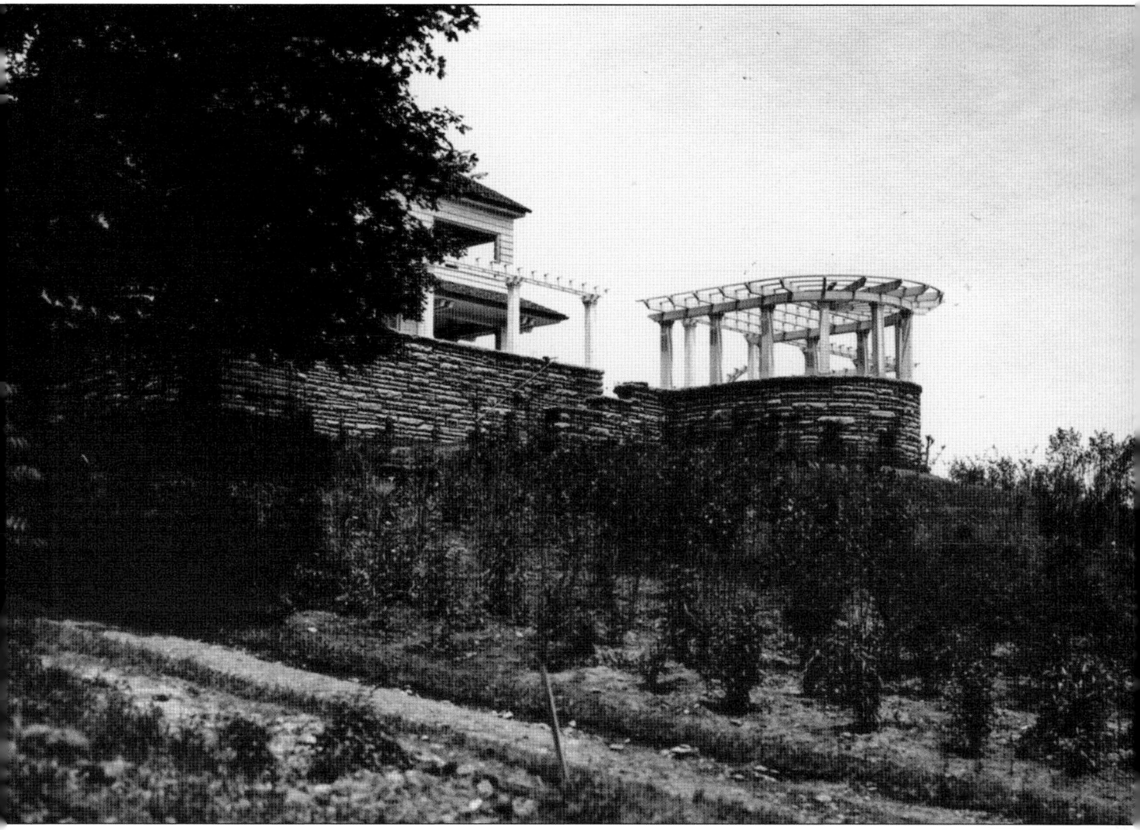

View of Western Elevation. In this image is a portion of the western elevation of the Wade's estate at Valley Ridge Farm, highlighting one of the mansion's stone promontories and intricately designed arbors. It also illustrates a typical Warren Manning design arrangement with the cluster of plantings to create a more natural layout. Remnants of this planting scheme can still be found at the original site. (Elizabeth Wade-Sedgwick.)

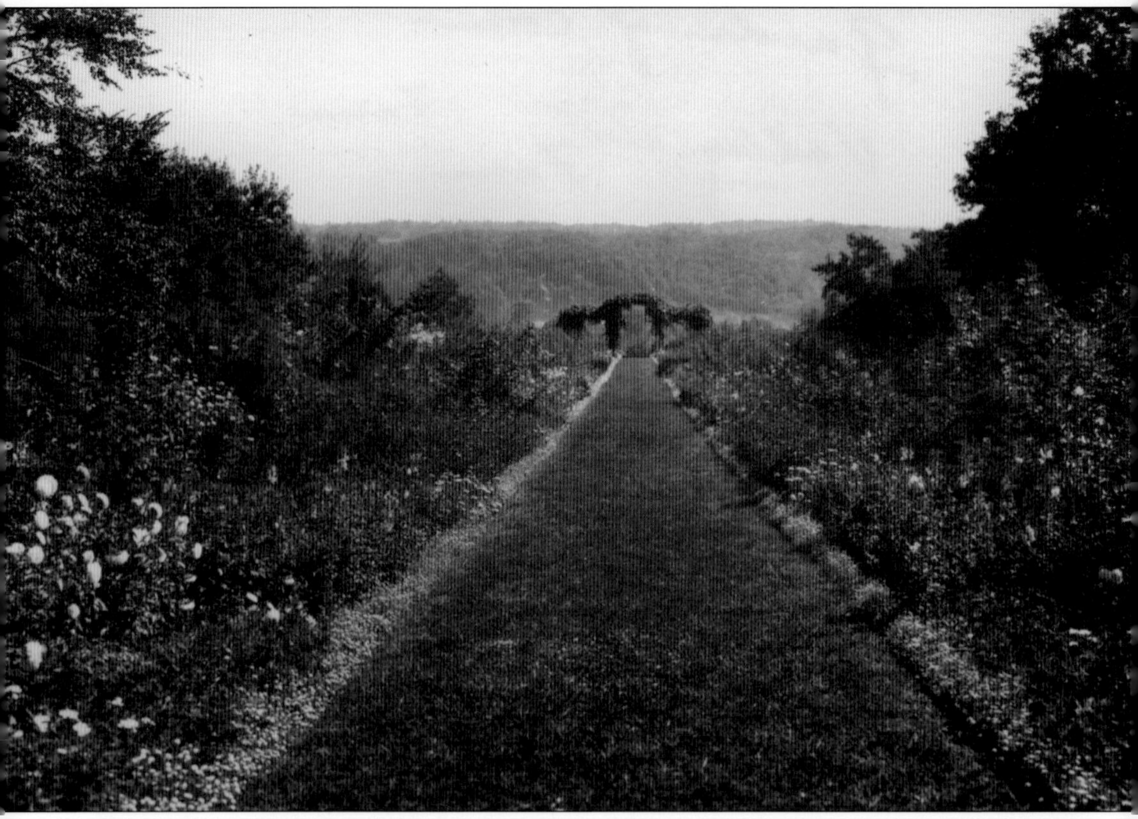

The Garden of Eden. This early photograph from the personal collection of the Wade family shows the Garden of Eden as Warren Manning intended it, complete with the picturesque hills in the background. Manning would carve out this patch of ground in the middle of the woods and designed it to contain all manner of fruit-bearing species of trees, shrubs, and vines. Standing by the gazebo, it was possible to peer down through one of the many vistas that had been cut to make visible either the Chagrin River to the east or the farm complex below. (Elizabeth Wade-Sedgwick.)

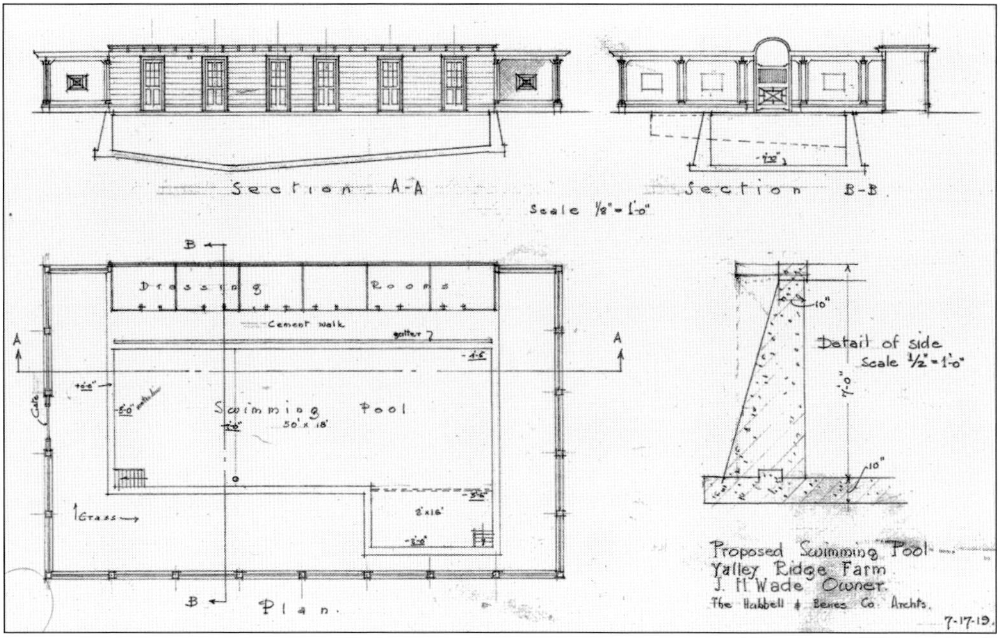

THE SWIMMING POOL DRAWING. Even the design of the swimming pool was meticulously detailed by Hubbell and Benes. Note the fence detail on the latticework. This is the signature design throughout the entire Valley Ridge Farm property. It would appear on all gates, fences, and railings. Even the butter produced at this farm would bear this symbol. (WRHS.)

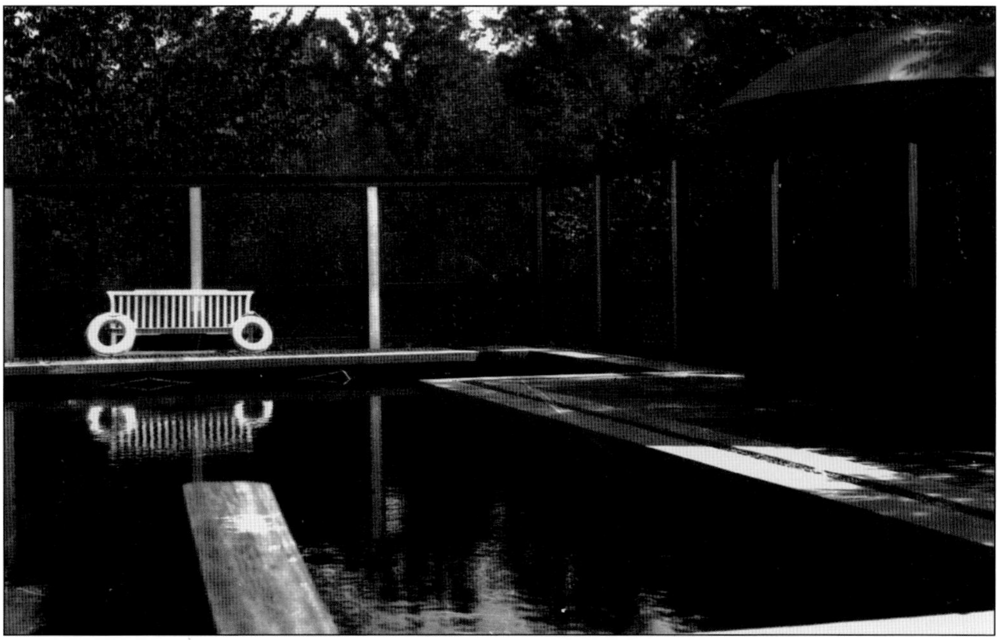

SWIMMING POOL. This photograph of the swimming pool on a summer's day shows the latticework and fence detail that surrounded it. Note the wooden diving board. The swimming pool, now in ruins, is on the neighboring Zukowski property. Given the dense forest growth surrounding it, it is not visible from any vantage point on university property. (Elizabeth Wade-Sedgwick.)

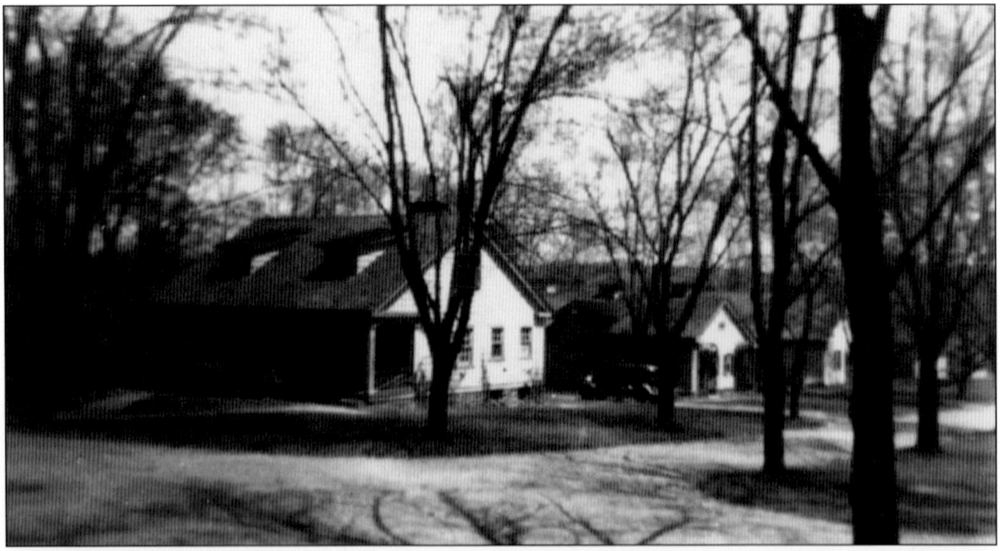

Caretaker's Cottage. This was intended to be the gardener's cottage, but there is no evidence to suggest that the gardener ever lived here. It would be home to the Wallace family for several decades and then in turn to numerous different families and individuals who worked the land at the farm until 2009. Note the design element on the front porch, which appears throughout the farm's buildings and fences to this day. (Wallace family.)

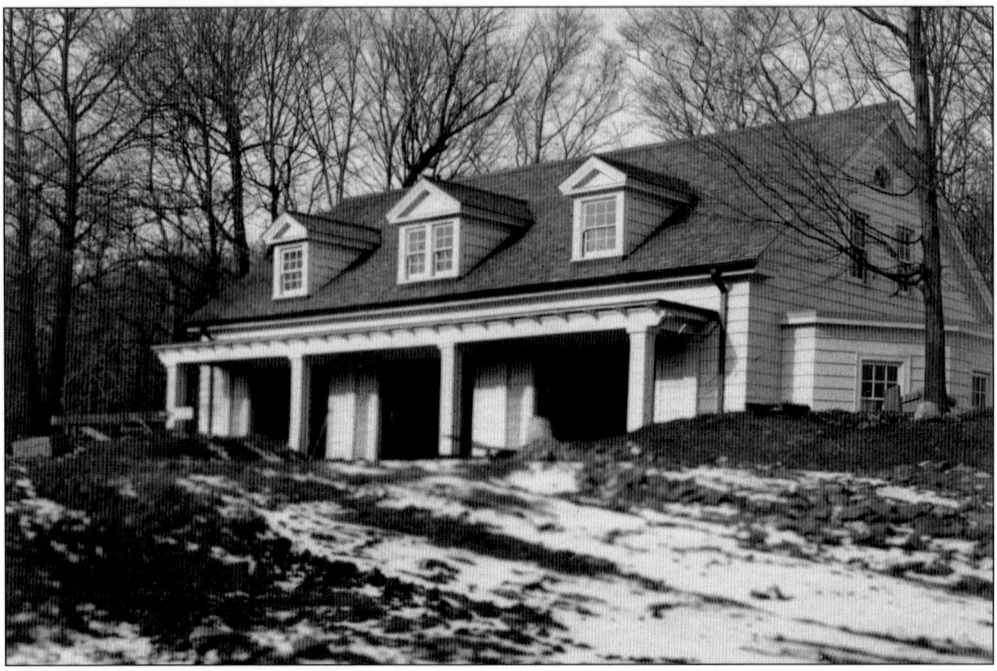

Superintendent's Cottage. This impressive home served as residence of the superintendents and their families at Valley Ridge for over 70 years. It has the trademark design element on its front porch. The cottage is now one of the private residences at University Farm. (Wallace family.)

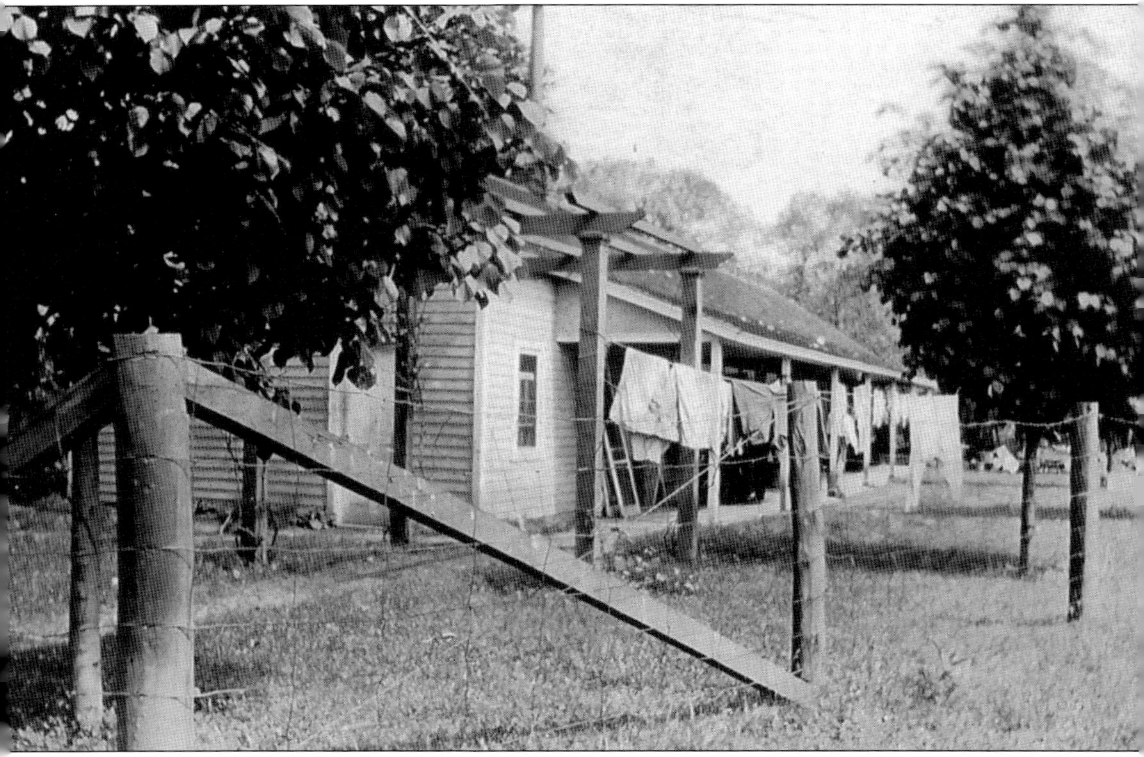

The Worker's Camp. Up to 50 men were employed seasonally by J.H. Wade II to assist in the smooth operation of his farm. These workers were housed on-site, either in this work camp or in quarters that were built in the upper portion of the horse barn. All types of glass bottles still turn up in the area surrounding the complex. Other artifacts that have turned up include newspapers, magazines, and packaging for a variety of products. Many of these are printed in various Eastern European languages. Cleveland has boasted the highest domestic population of Slovenes, Hungarians, and Poles within the United States, and it is no surprise that many men with these backgrounds would find their way to work at Wade's farm. (Wallace family.)

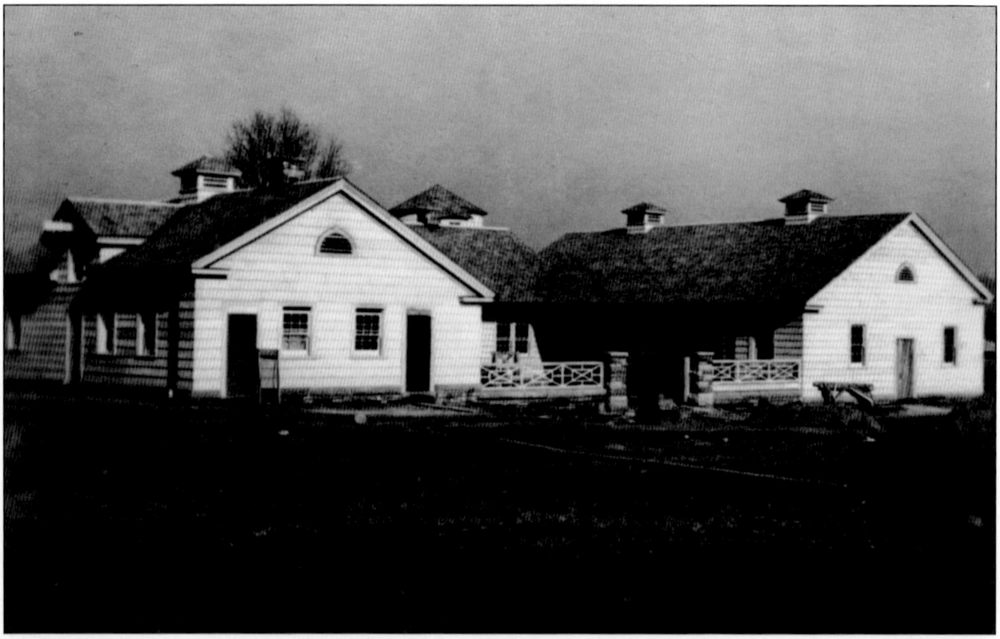

The Dairy Barn. This barn was the envy of Wade's dairy farming peers, and it would be the envy of any dairy farmer today. It was complete with the following: a fenced-in bull yard; an attached silo; calving rooms for expectant bovine mothers; separate quarters for calves, bulls, and cows; scales of all kinds for weighing the animals and the milk; and an attached room by the ice room for keeping the milk and butter cold. (Wallace family.)

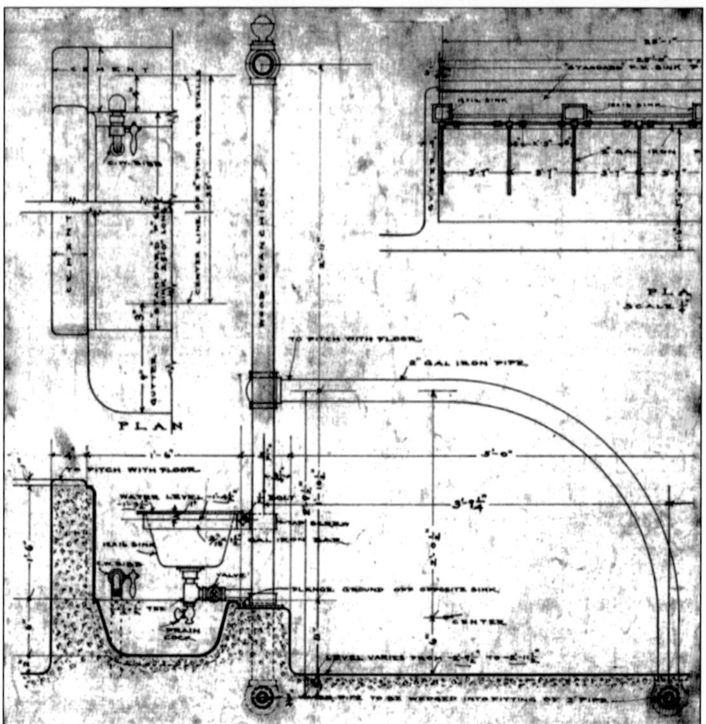

Drawing of the Stanchions. This drawing by Hubbell and Benes details the stanchions for milking the cows. These have been restored and can be viewed in their original location. Their size suggests that smaller breeds were milked in them, as many of today's breeds would be too large to fit. (WRHS.)

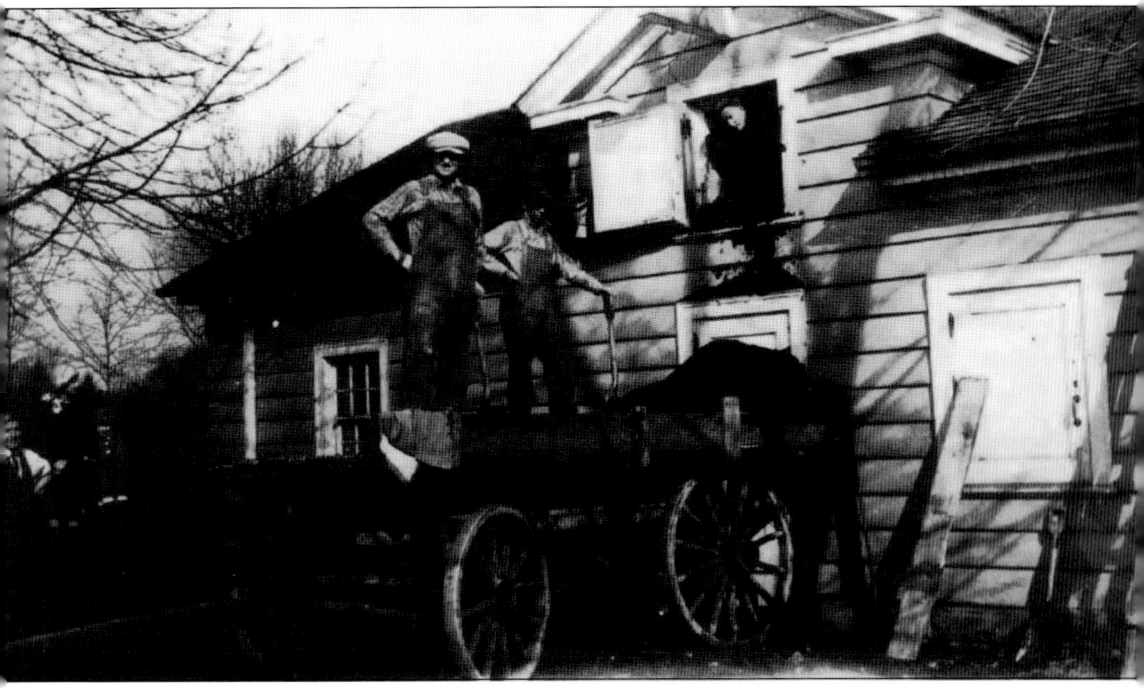

The Ice Room. One of the more interesting features of the Wade summer estate was the large room attached to the dairy barn that served as an ice repository. A man-made lake across the field served as the source for large ice blocks that were cut in the winter, hauled across the field, and stored in the ice room. This allowed for refrigeration throughout a large portion of the year for the dairy products that were produced at Valley Ridge Farm. Here, the men are seen putting sawdust in the ice room to act as insulation. On the other side of the ice room is a built-in c. 1910 refrigerator that was kept cool by its proximity to the ice room. (Wallace family.)

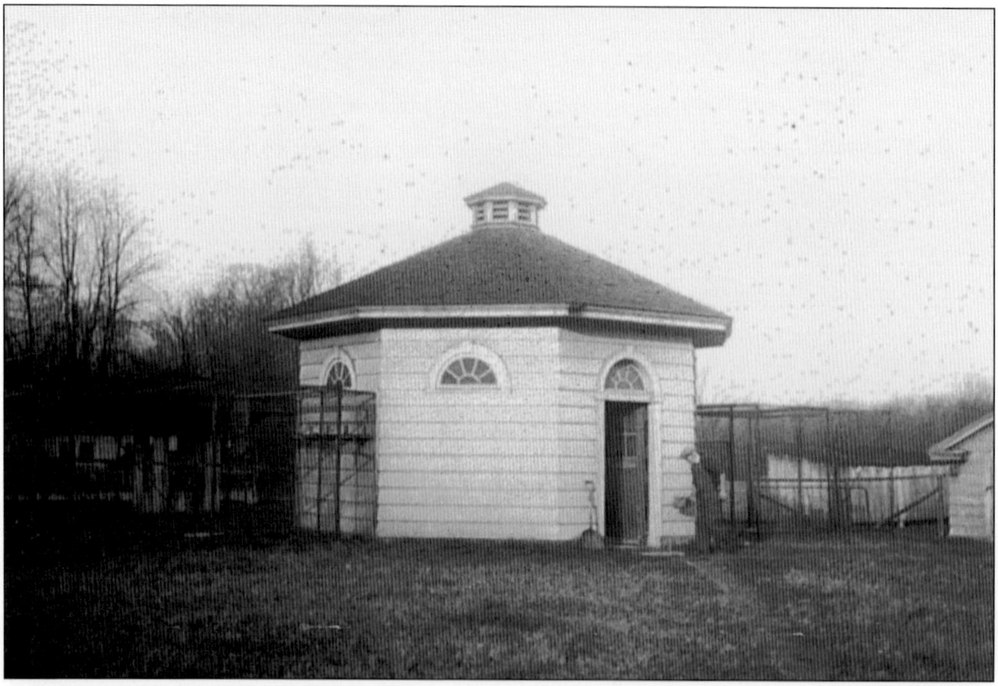

The Dovecote. The dovecote, or aviary, was where birds (namely doves) were raised to be served at either Wade's Cleveland home or enjoyed right here at his farm. Today, this restored building houses supplies for the many students who take art classes at Valley Ridge Farm. (Wallace family.)

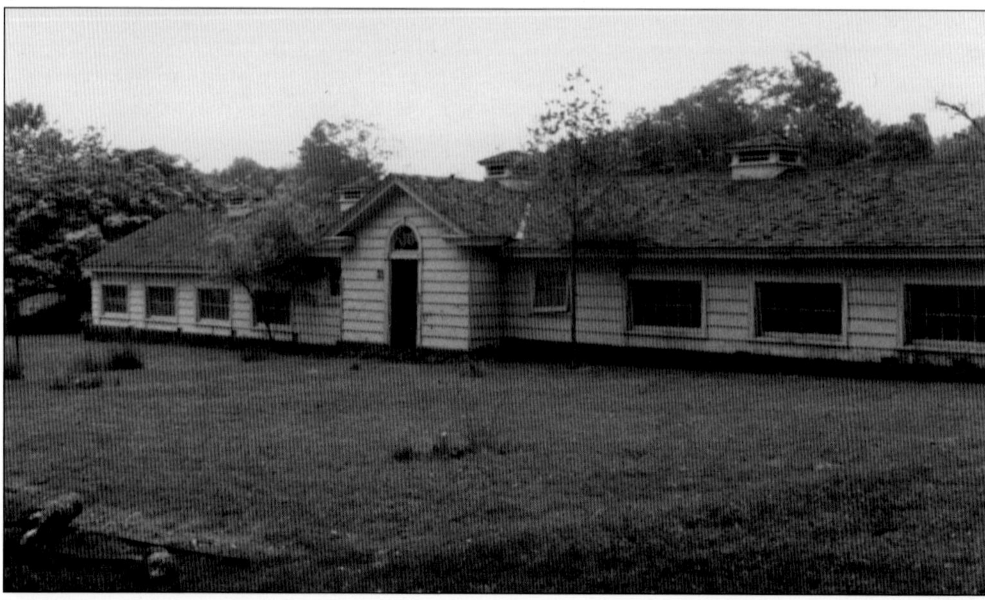

The Chicken Coop. No less impressive in its day than any of the other buildings at Valley Ridge Farm was the chicken coop. Its user-friendly design allowed the caretakers to tend to the chickens via a hallway with several access points so that they would not have to enter the individual rooms and disrupt the chickens unless necessary for tasks, such as cleaning. (Wallace family.)

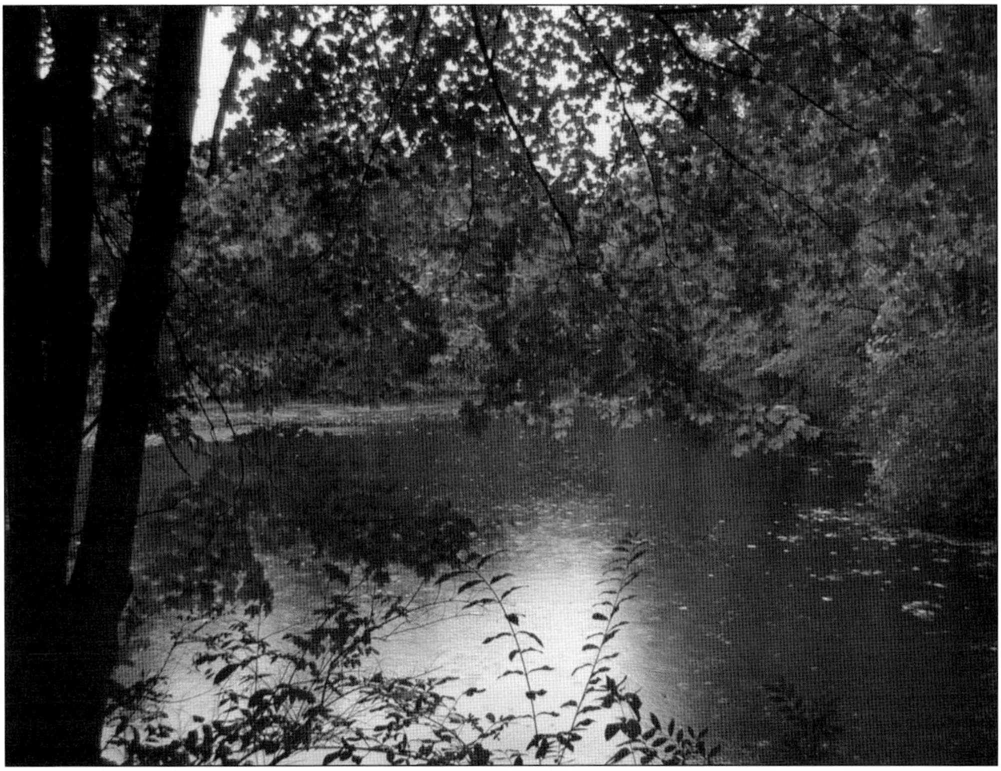

The Fishing Pond. The fishing was said to be good at the fishing hole on the Wade estate. This pond is still accessible and can be seen while taking the walking path from the upper Squire Valleevue Farm down to the lower Valley Ridge Farm. A windmill sited in the nearby restored prairie area provides aeration to the pond. It is also the site of many of the community education and school group visitation programs performed by farm staff members.

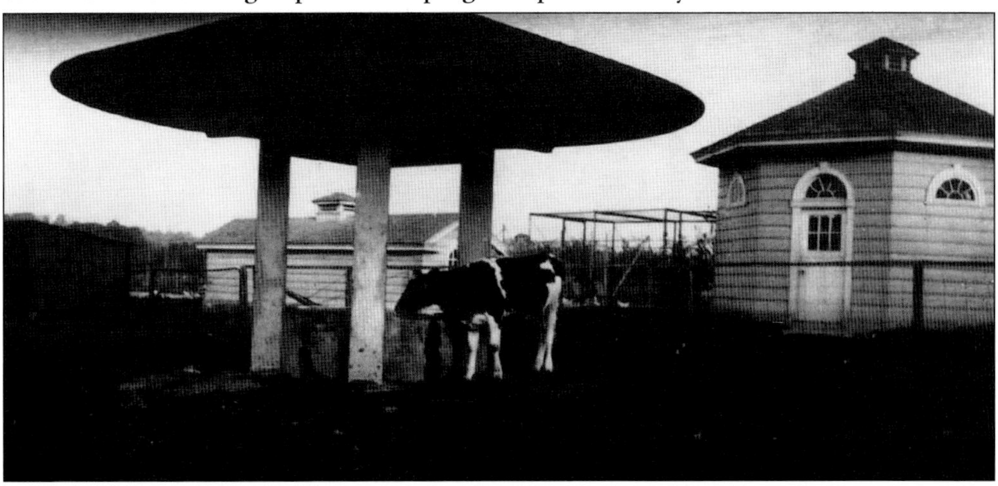

Water Trough. Scattered throughout the landscape were a few dome-roofed water troughs, such as this one. They were designed by Hubbell and Benes to complement the rest of the compound's architecture. The livestock at Valley Ridge Farm had some of the most handsome watering holes in the entire village. Two of these structures can still be seen today. (Wallace family.)

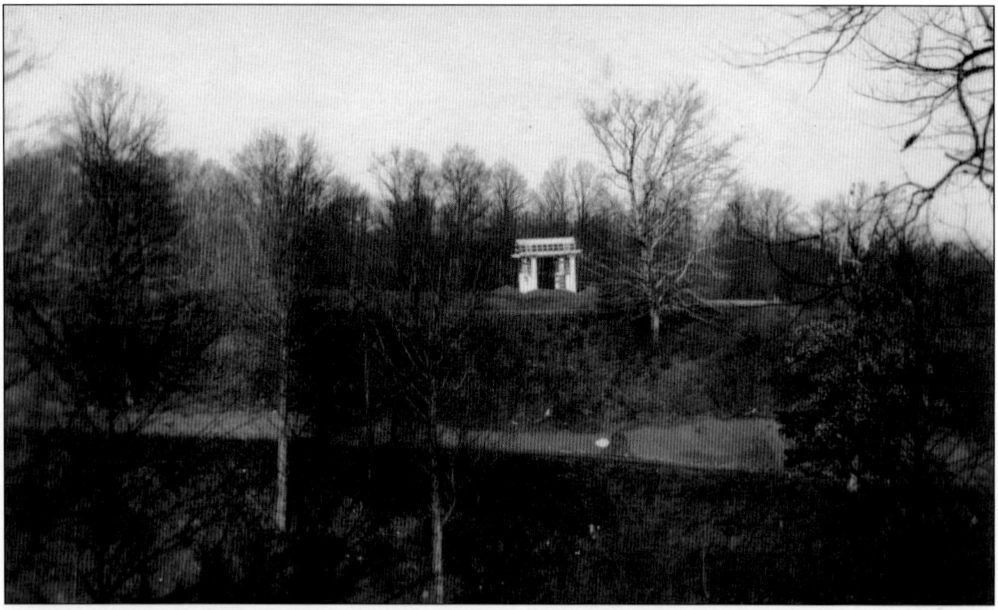

Gazebo at Valley Ridge Farm. At least three ornate gazebos were designed and installed among Manning's gardens, all bearing the signature design element of Valley Ridge Farm (the Garden of Eden gazebo is pictured above). Two that are located on the upper ridge are currently in ruins and awaiting reconstruction and renovation; one that is the centerpiece of the grape arbor (below) was successfully restored in 2009. Their location can all be seen on Manning's 1910 landscape drawing. Majority funding for the grape arbor restoration has been kindly provided to the farm by Ray and Mary Ann Leonard Zukowski. (Above, Elizabeth Wade-Sedgwick.)

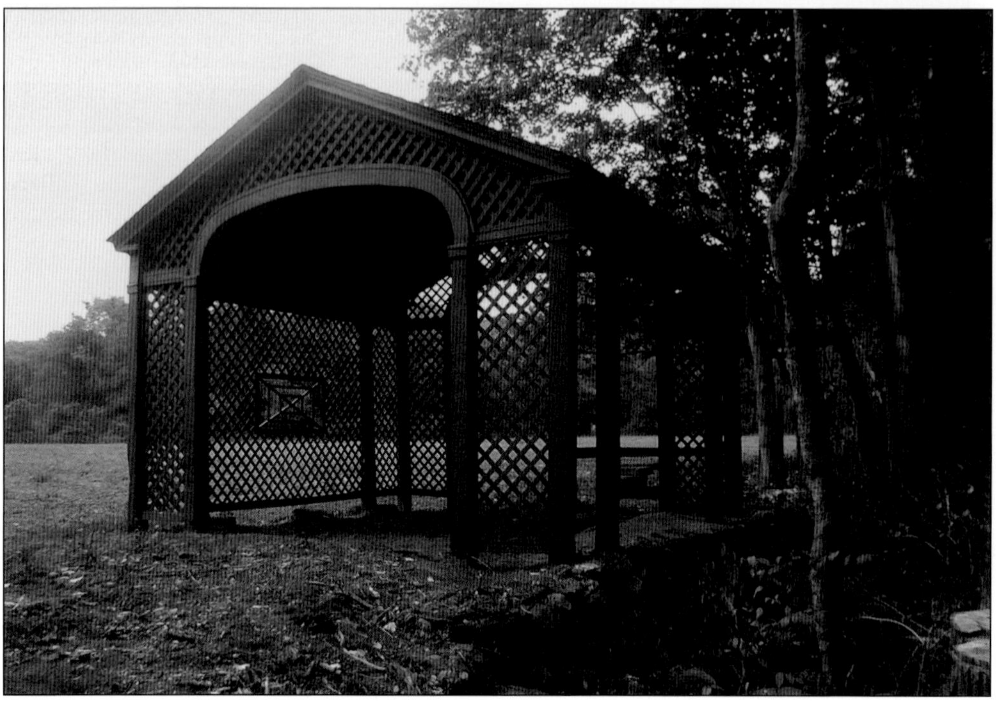

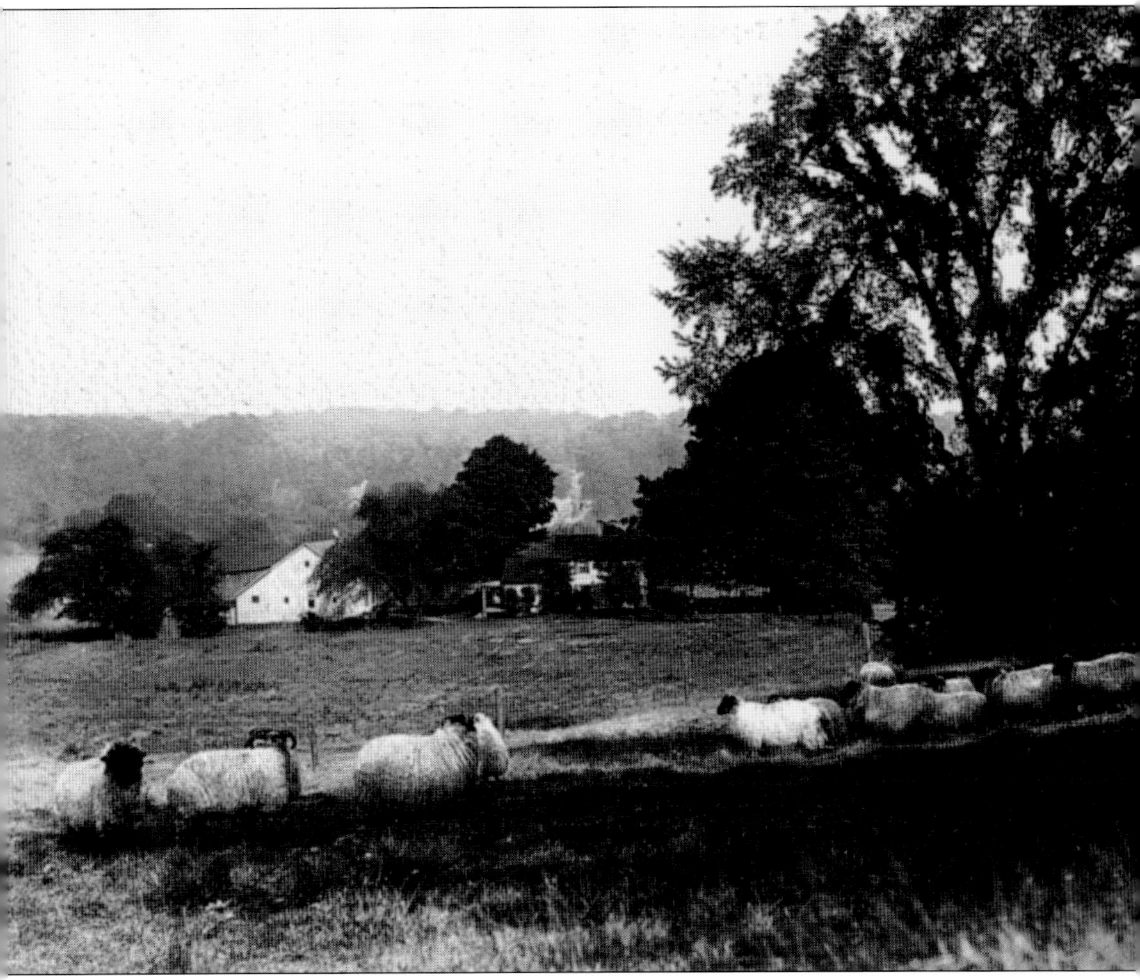

Sheep Grazing. Pictured is a flock of sheep, both rams and ewes, grazing in one of the Wades' pastures. The responsibility of overseeing the sheep would be undertaken by the Wallace family, who would remain in the employ of the Wades for many decades. The entire design of Valley Ridge Farm was meticulously planned out by both Warren Manning and Hubbell and Benes, including these grazing areas. Not a detail was left out. (Wallace family.)

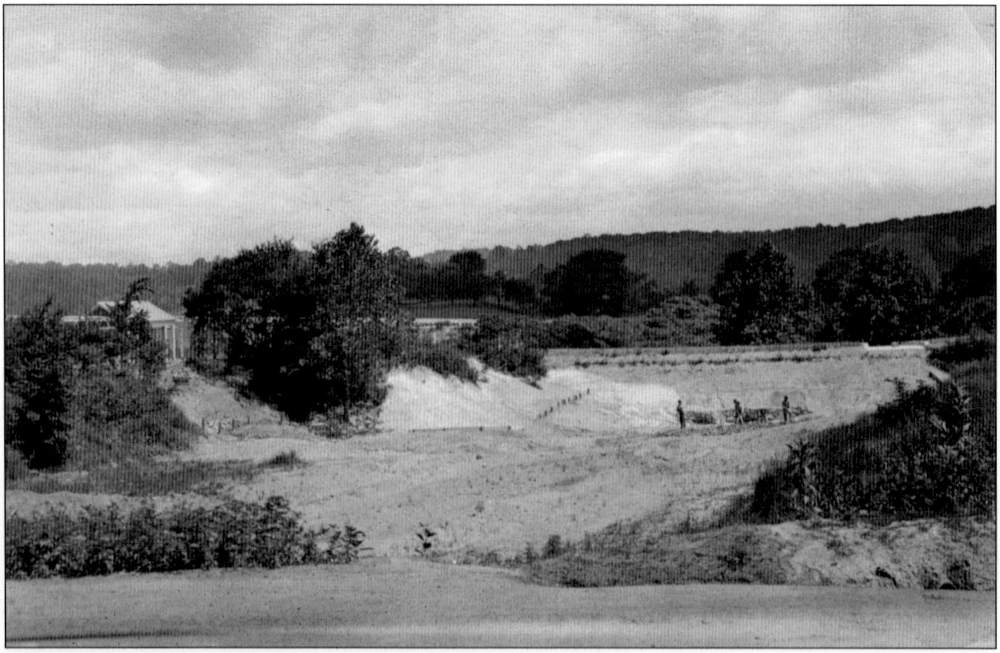

Building the Dam. This 1910 photograph shows the dam being constructed in the recently excavated area that was to be the lake. This is the same man-made lake that would be the source of ice blocks to fill the ice room. The grape arbor and gazebo can be seen along the northern ridge of the lake. (Elizabeth Wade-Sedgwick.)

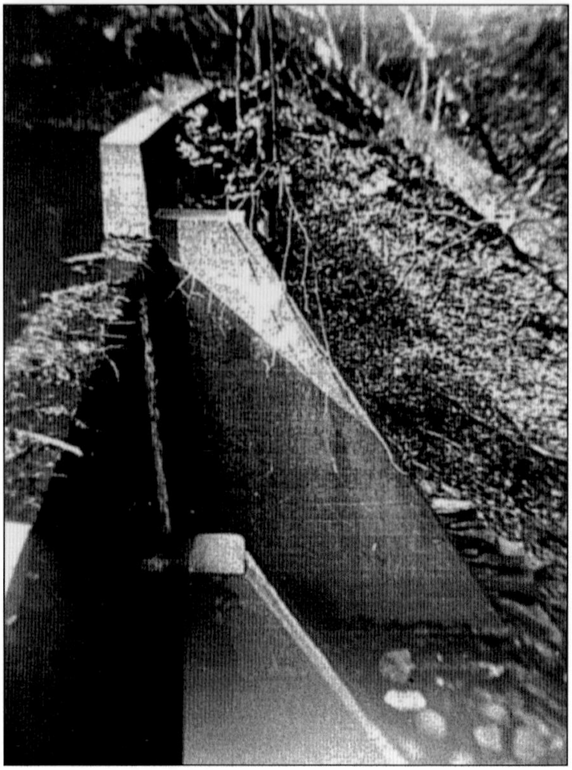

View of the Dam. This was the dam that retained water in the one-acre lake that was the source of the blocks of ice. It was located just south of the truck garden fields. Eventually, time and gravity took its toll, and the dam collapsed. The majority of it still stands, and the ruins of the rest can be seen in the valley below. Trees and growth now occupy the void where the lake once was. (Wallace family.)

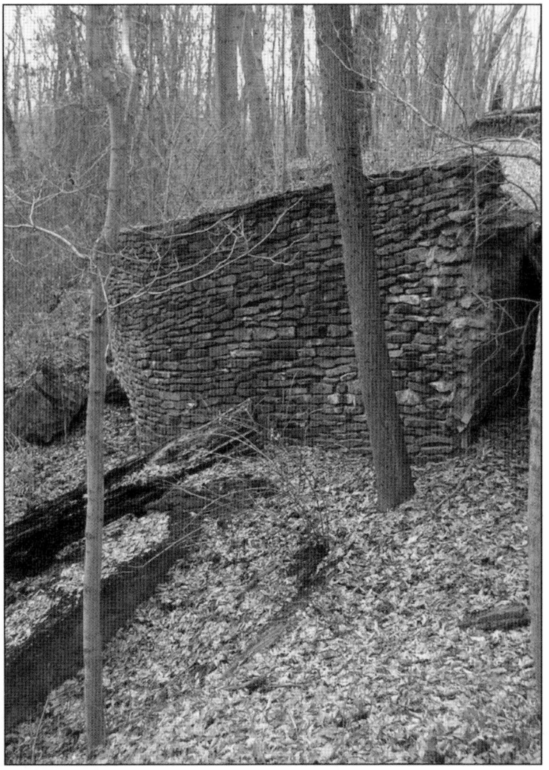

Stone Retaining Wall. Though slightly worse for wear, this 100-year-old retaining wall remains in full service, overlooking one of the vales on the upper portion of the farm. It can still be seen on the walking path from the upper farm to the lower farm and is just east of the Garden of Eden.

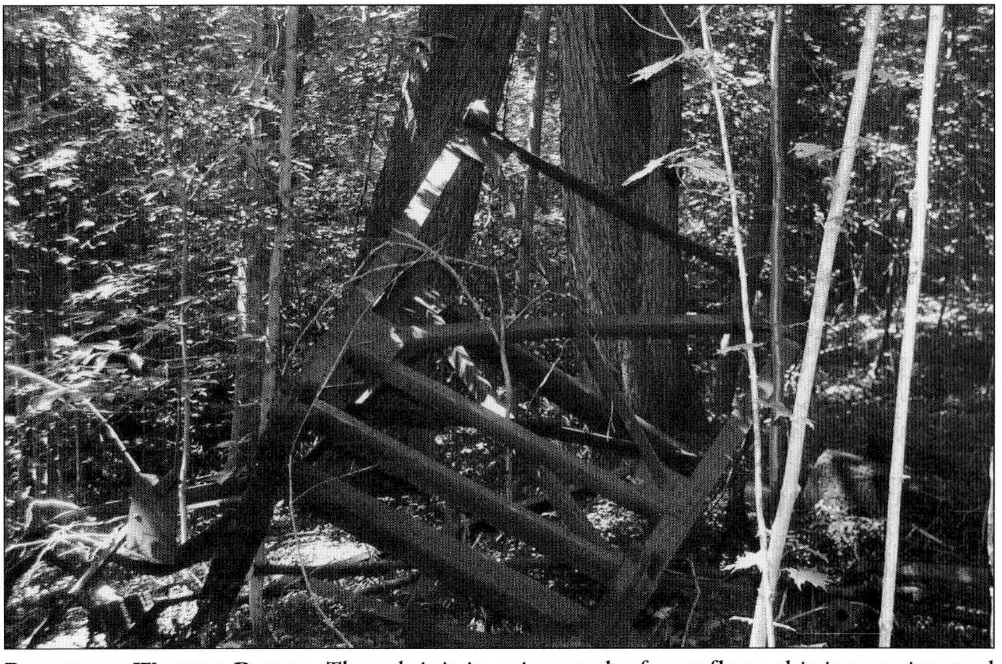

Pedestrian Walking Bridge. Though it is in ruins on the forest floor, this impressive steel structure, which now looks like a modern art sculpture, can still be seen at its former site. It connected the mansion to the Garden of Eden, crossing over the deep ravine that divided two portions of Wade's summer estate.

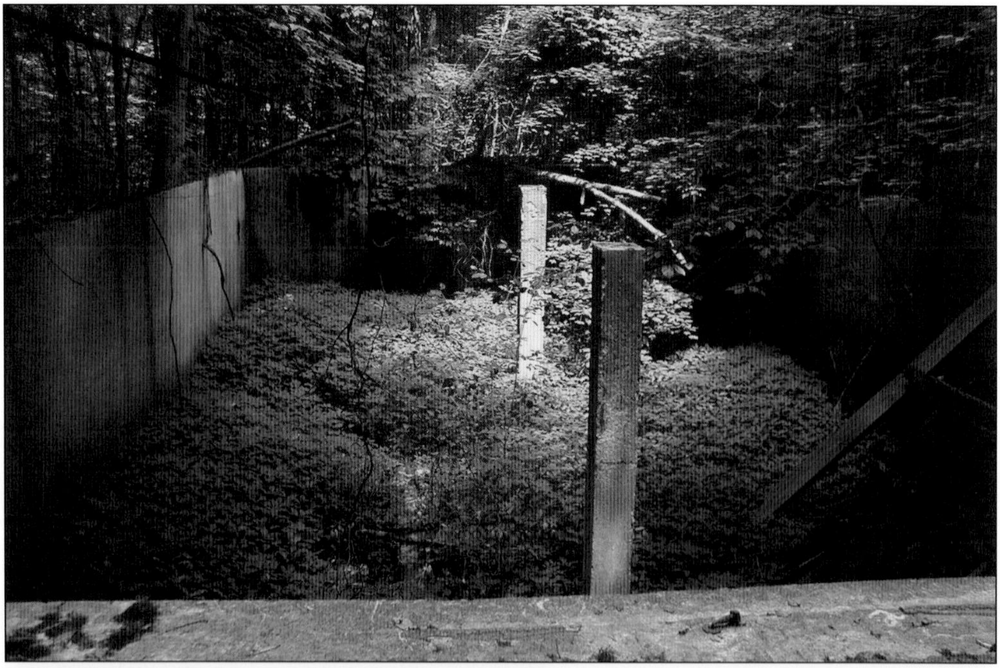

The Water Reservoir. J.H. Wade II's farm complex was fully plumbed with running water, an impressive feat for the early 20th century in this rural area. This 115,000-gallon reservoir located on the upper portion of his land would supply water through a series of pump houses to the farm and buildings below.

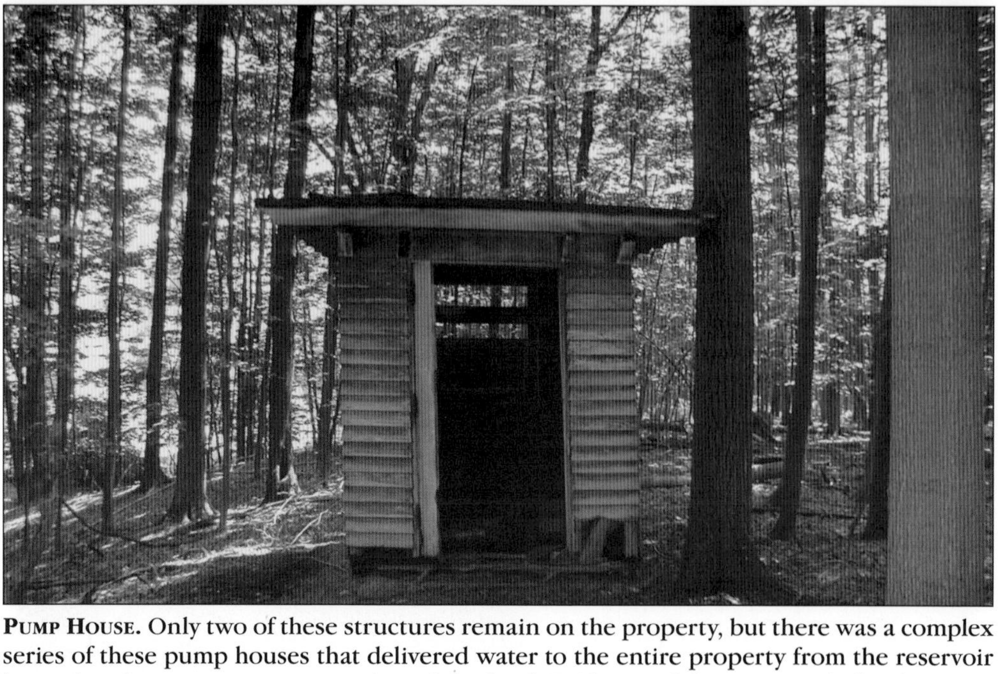

Pump House. Only two of these structures remain on the property, but there was a complex series of these pump houses that delivered water to the entire property from the reservoir located at the upper western portion of the land. Evidence of the pipes and plumbing can still be seen throughout the property. A similar looking hose house, which served to water the vegetable gardens, still exists and was restored between 2009 and 2010.

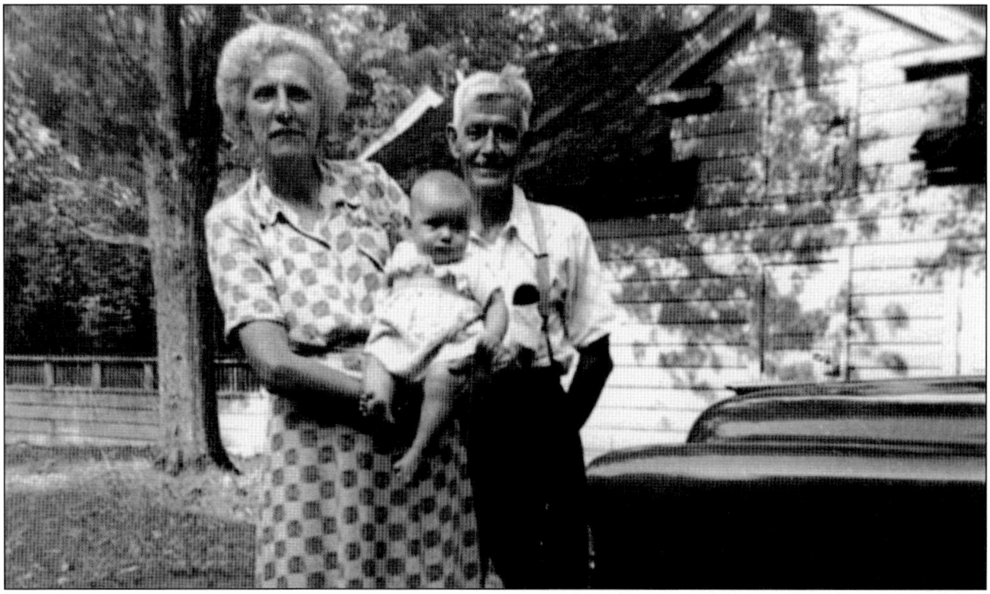

The Wallace Family. For four decades, the Wallace family was employed by the Wades as superintendents and stewards of the farm. They tended to the livestock, gardens, and upkeep of all the buildings. The personal family photographs of the Wallaces and their generosity in allowing University Farm to use the images has been responsible for much of the understanding of the history of the farm and its day-to-day operations during a large part of the 20th century. (Wallace family.)

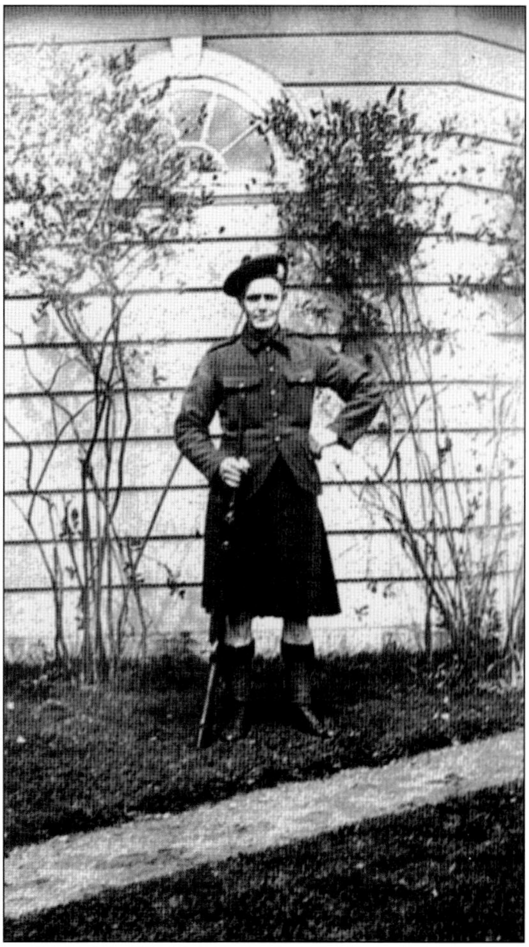

William Wallace. Wallace family patriarch William Wallace is pictured on his wedding day. Wallace was an immigrant from Scotland who would work at Valley Ridge Farm from 1917 until 1957 as caretaker of the farm's flock of sheep. William and his family lived in the caretaker's cottage situated right next to the dairy barn and across from the chicken and horse barns. (Wallace family.)

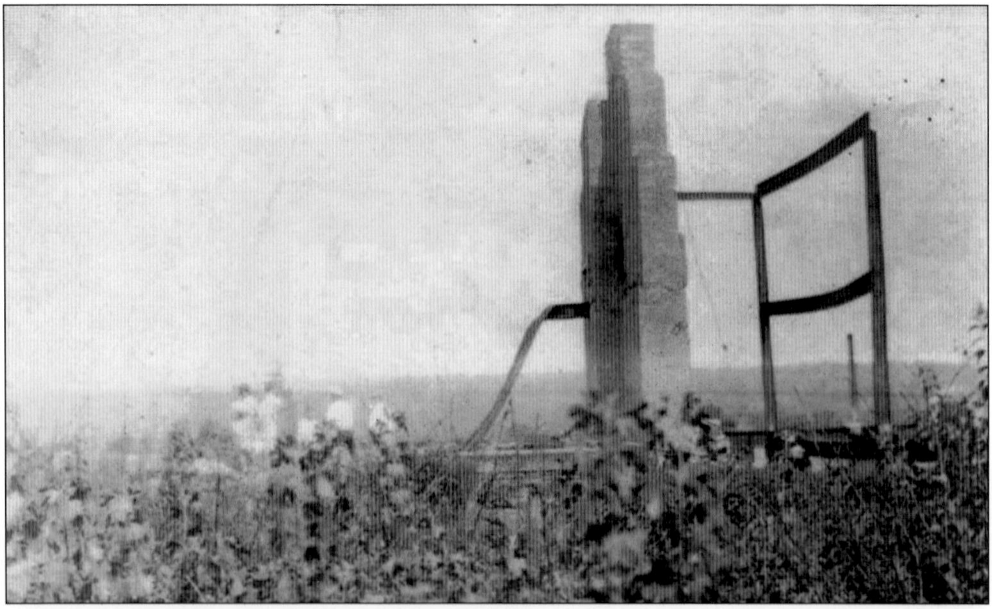

AFTER THE FIRE. The Wade's Manor House was consumed by fire in the winter of 1929. Oily rags left carelessly by painters during the house's annual maintenance were the cause. It is said that $10,000 dollars in artwork (in 1929 dollars) burned off of those walls, including an original Winslow Homer painting. (Wallace family.)

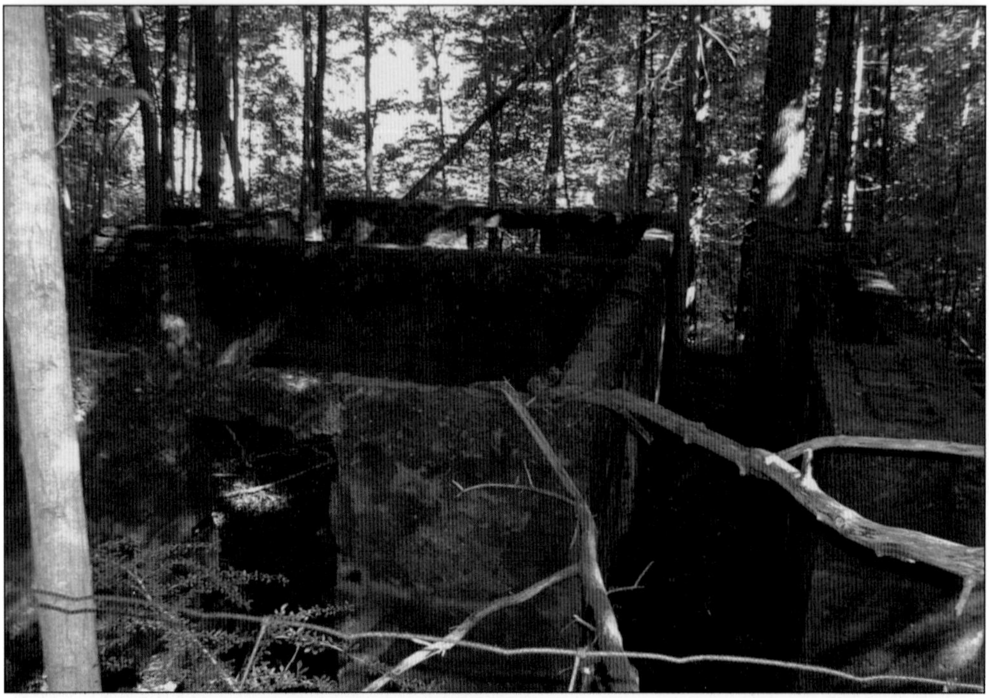

THE RUINS AS SEEN TODAY. Access to the site of the former Wade Manor House is still possible, but only by traversing a steep hill via an easement, where mostly only deer tread. One can get a sense of the prior grandeur by surveying the footprint of the foundation. The hollow expanse is now overgrown with opportunistic trees, shrubs, and undergrowth.

Five

Restoration of Valley Ridge Farm Facilities 1977–2009

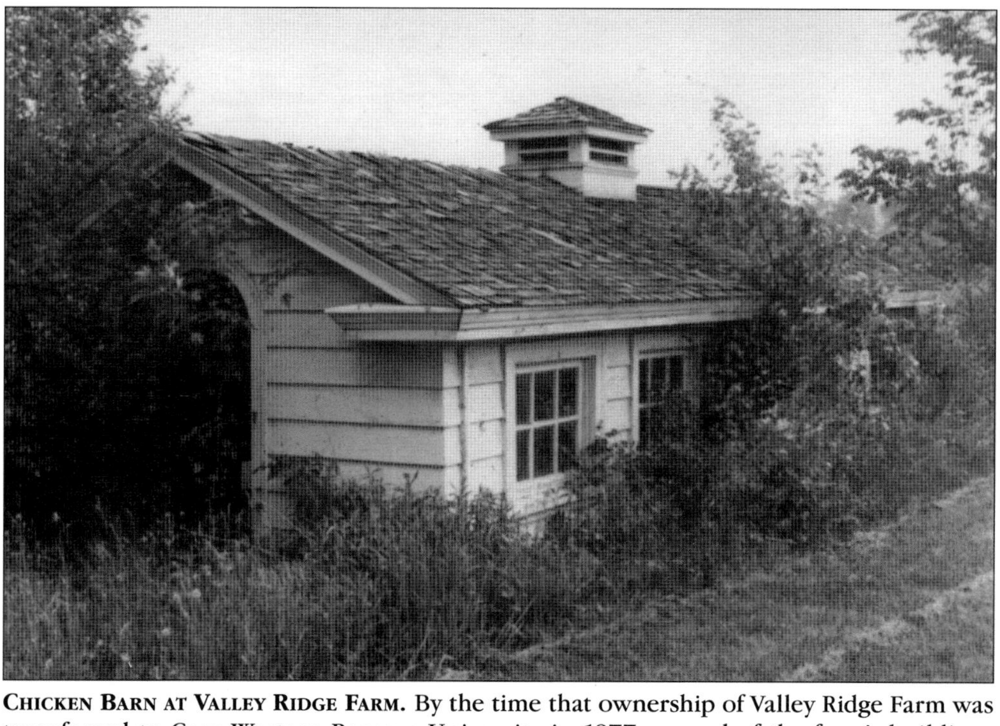

Chicken Barn at Valley Ridge Farm. By the time that ownership of Valley Ridge Farm was transferred to Case Western Reserve University in 1977, several of the farm's buildings had suffered the ravages of vines and overgrowth due to decades of nonuse. (Elizabeth Wade-Sedgwick.)

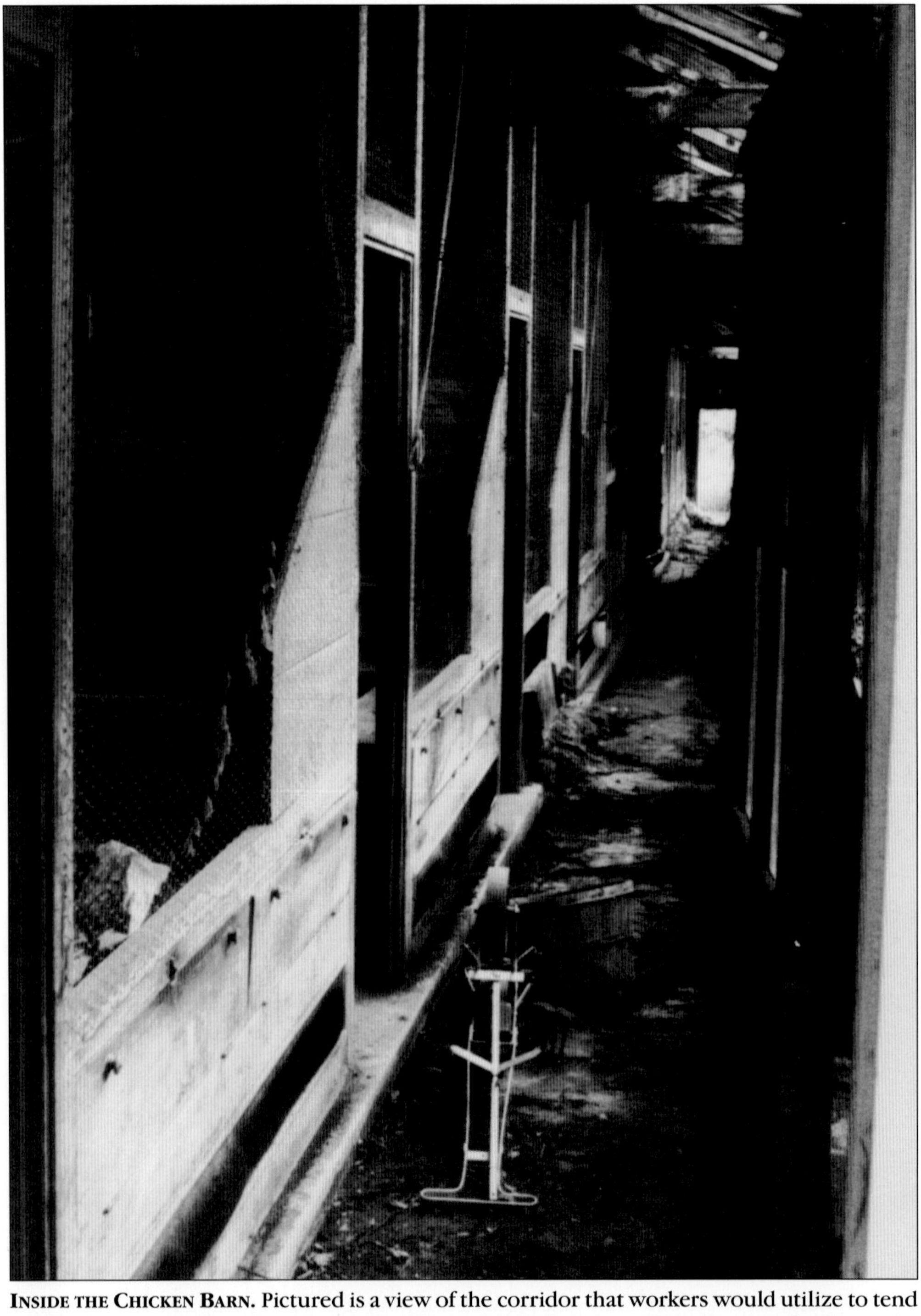

INSIDE THE CHICKEN BARN. Pictured is a view of the corridor that workers would utilize to tend to the chickens without having to enter the individual rooms. This photograph shows the interior many years after it had last been used to keep chickens. Though the poor condition of the building is not evident in this image, the exterior of the building was deemed irreparable, and it was burned down. (Elizabeth Wade-Sedgwick.)

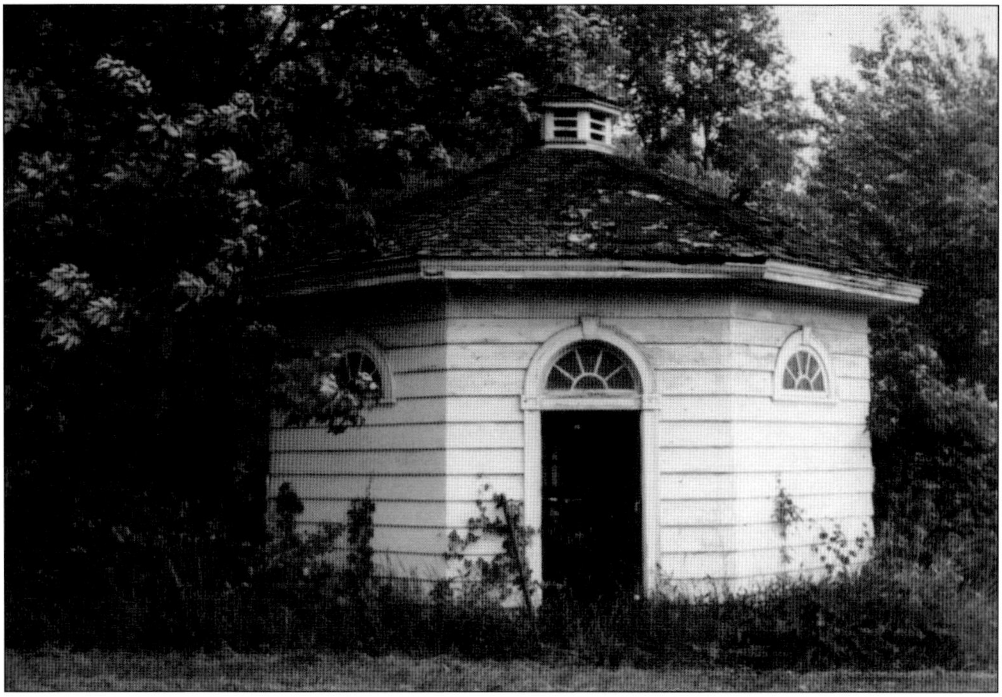

Dilapidated Structures at Valley Ridge Farm. These photographs illustrate the conditions that the farm buildings were in before renovations commenced. In many cases, nuisance animals, such as raccoons, had free range to live for many generations in the upper stories. This created a challenge for farm staff who had to clean them out. The dovecote can be seen above, and the water trough can be seen below, with the pig barn poking out from behind but barely visible with the festoons of wild vines draping off of it. (Both, Elizabeth Wade-Sedgwick.)

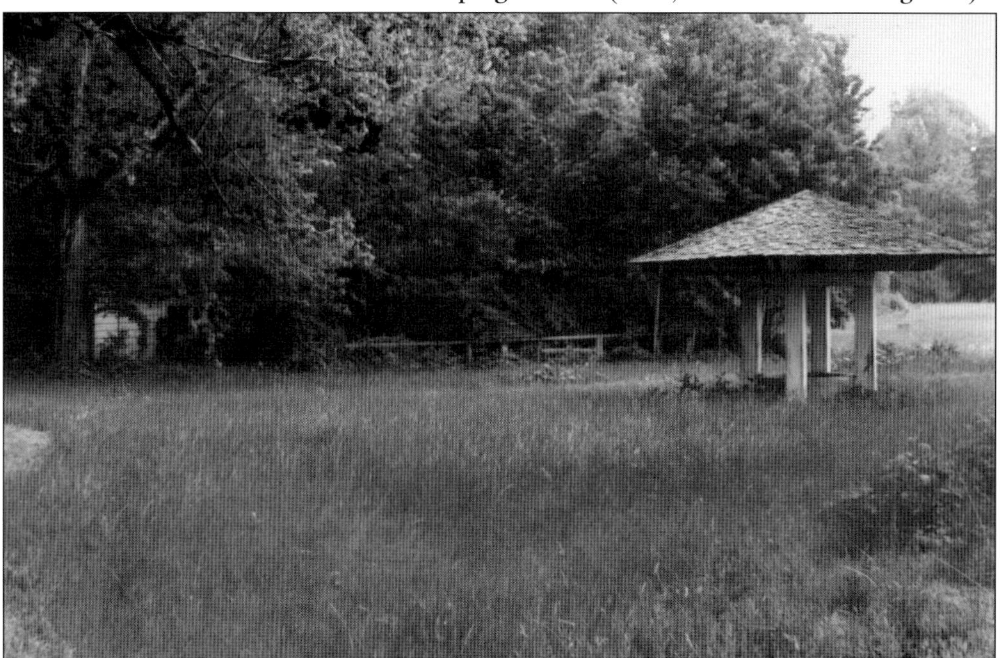

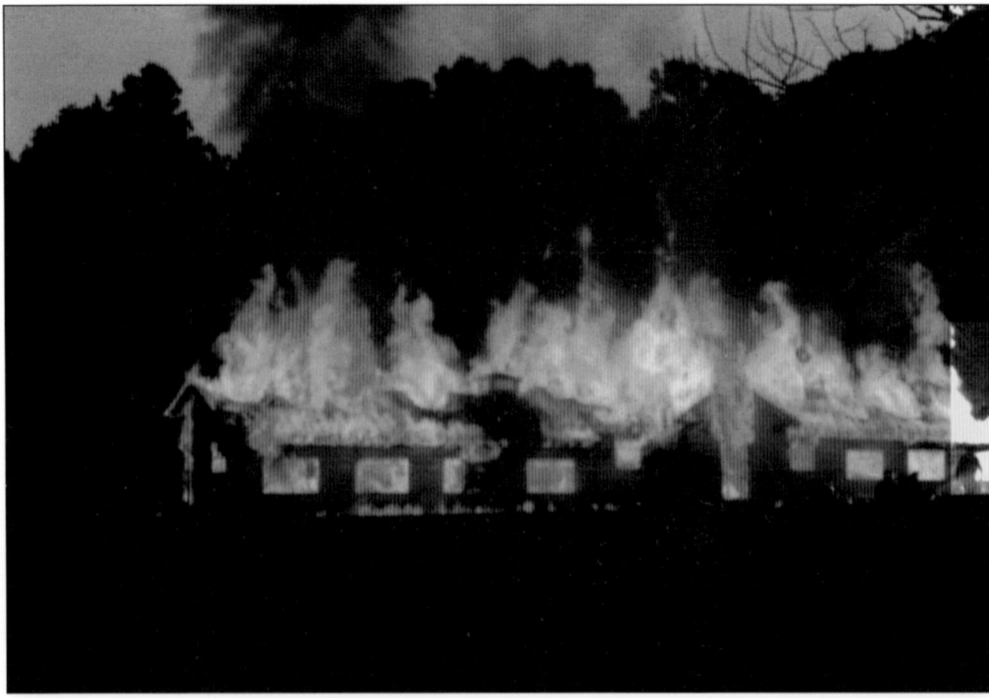

Chicken Barn Burned Down. Decisions were made building by building on their respective fates. Most were fortunately deemed salvageable and usable; a few, however, were not. The chicken barn was determined to be far too damaged to be worth the effort, and with the assistance of the local fire department, it was burned to the ground.

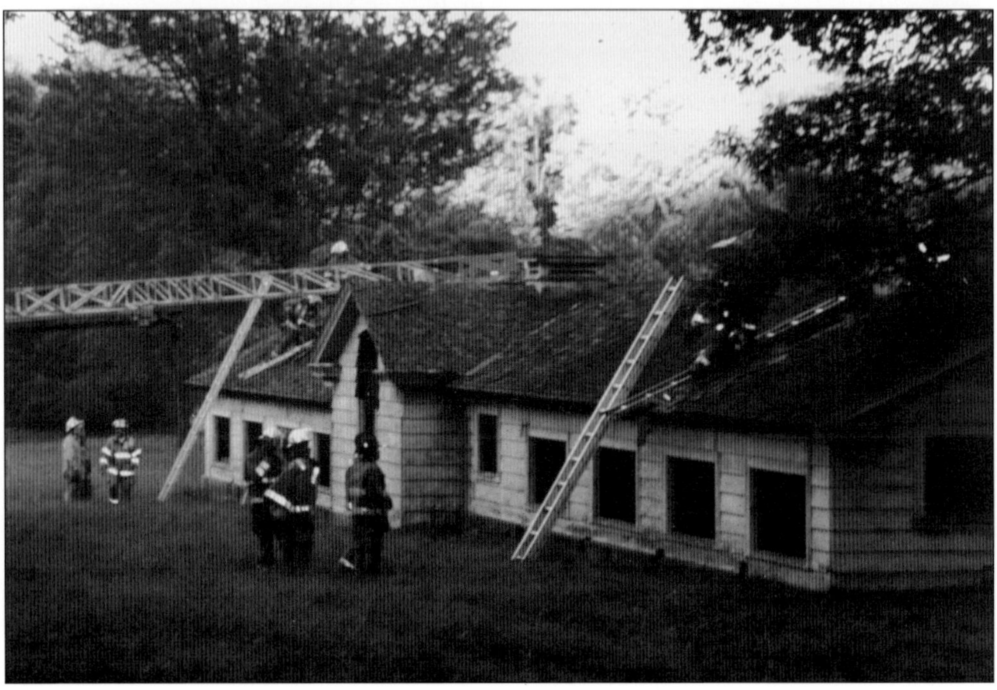

Pig Barn. This building now serves as the ceramics studio, but by the time the university took possession of the pig barn, it was overgrown with vines and brush and in need of a new roof, new electric service, and foundation work. Most of the unique features were retained, including the pig doors, which were operated by pulley from a hallway inside.

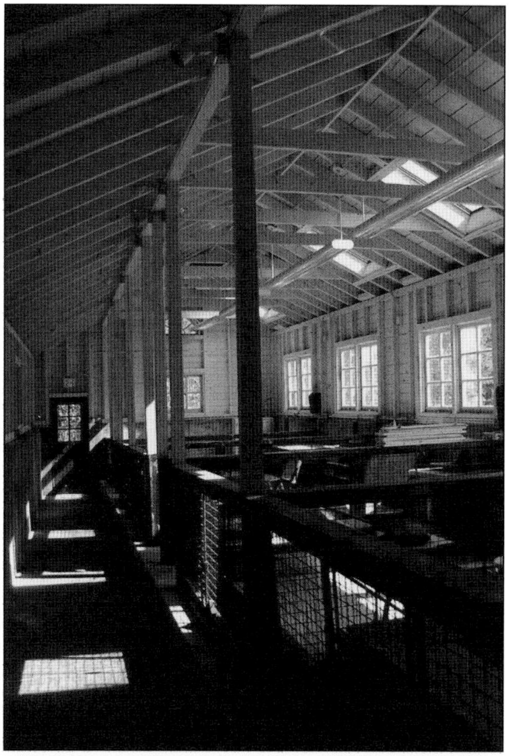

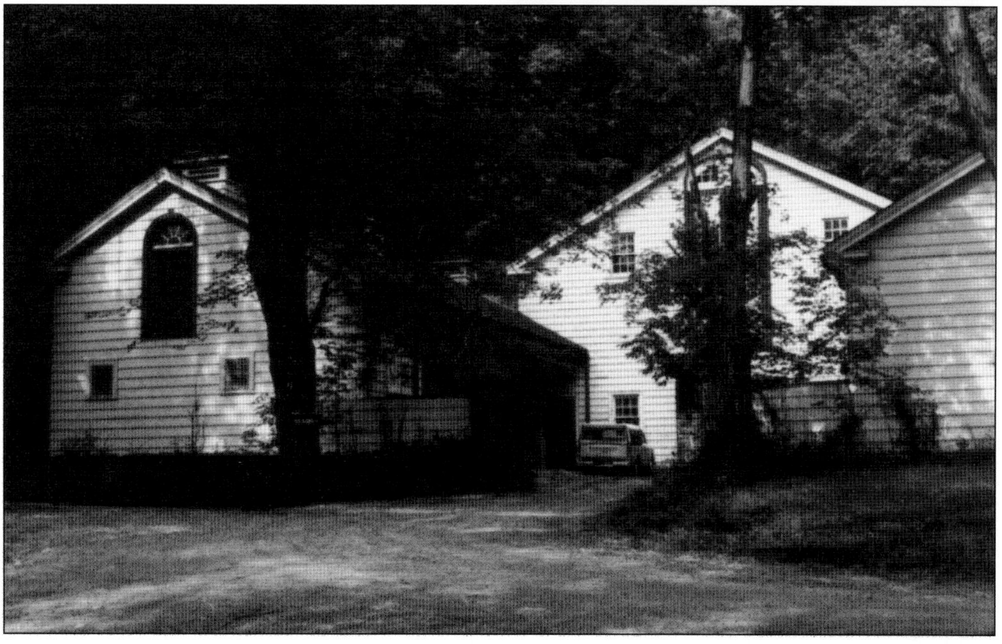

Horse Barn prior to Renovations. Designed by the firm of Hubbell and Benes, this barn was said to be the most charming structure in all of Hunting Valley in the early part of the 20th century. Decades of neglect nearly destroyed it. Hubbell and Benes also built Wade's estate in Georgia, the Cleveland Museum of Art (on land donated by Wade), and the famed Wade Chapel in Lakeview Cemetery, complete with Tiffany windows.

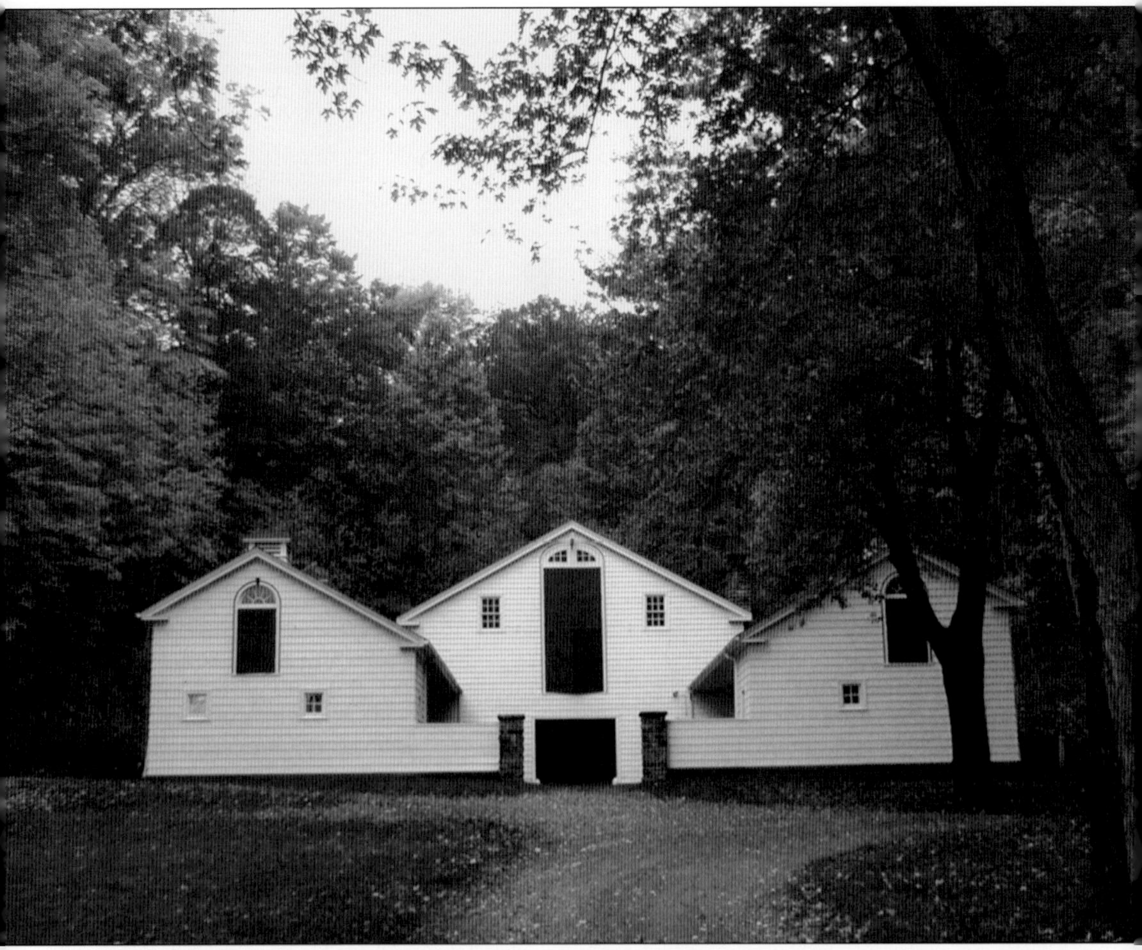

RENOVATED HORSE BARN. The horse barn at Valley Ridge Farm was slated for demolition before it was saved with the help of many devoted individuals who had a desire to preserve history and donated funds. Over many decades, the hills around the horse barn had eroded to the point where soil covered much of the foundation, which caused serious rotting of the original wood siding. Heirs of Jeptha Homer Wade II would generously contribute the majority of the money to start the restoration of the largest structure (at 11,290 square feet) on this portion of University Farm.

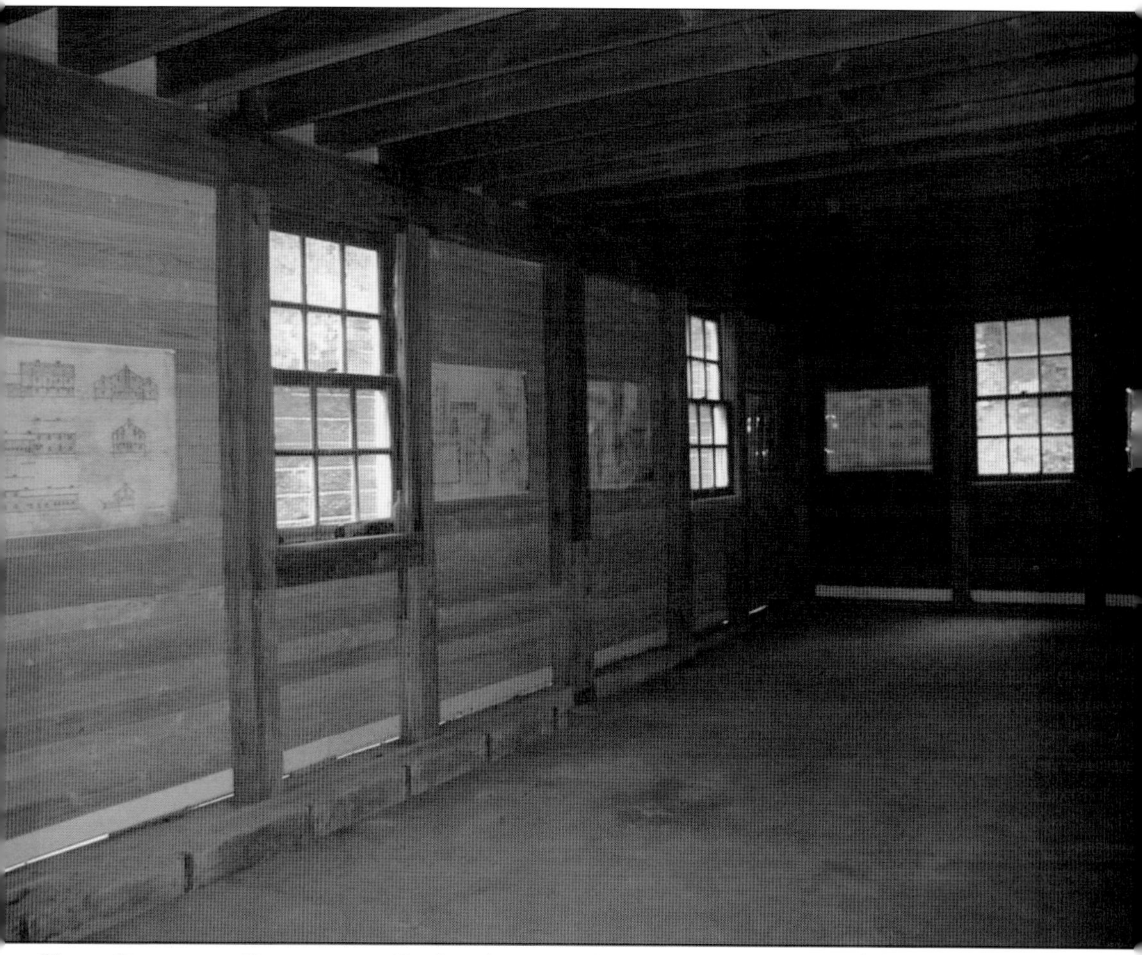

Horse Barn after Renovations. During the extensive renovation process of the horse barn, many of the contractors involved would note that this structure was a barn in name and purpose only, as the quality of workmanship, layout, and materials used were far superior to most homes built then or now. Visitors to the horse barn today still marvel at its grandeur. With the aid and assistance of the donors and skilled craftsmen, a piece of history was able to be saved for generations to come. A display in the interior of the barn highlights many aspects of its renovation process.

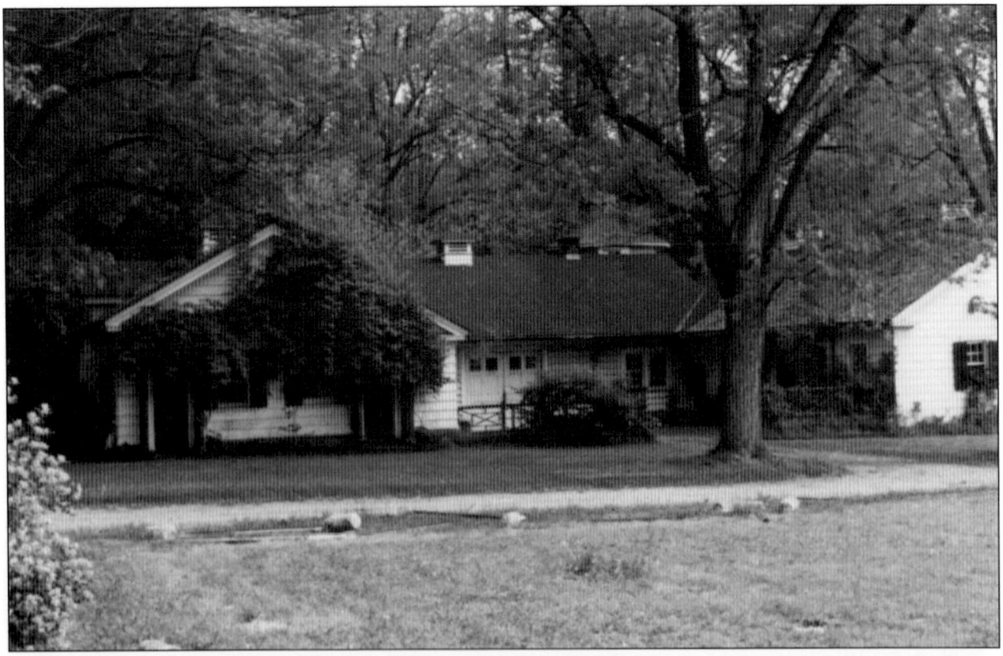

Dairy Barn prior to Restoration. The second largest structure on the property was also in need of major renovations. Time had ravaged many parts of it, including ceilings and walls. Fortunately, help would be forthcoming. The first major renovation came at the beginning of the new millennium, and the second was in 2009.

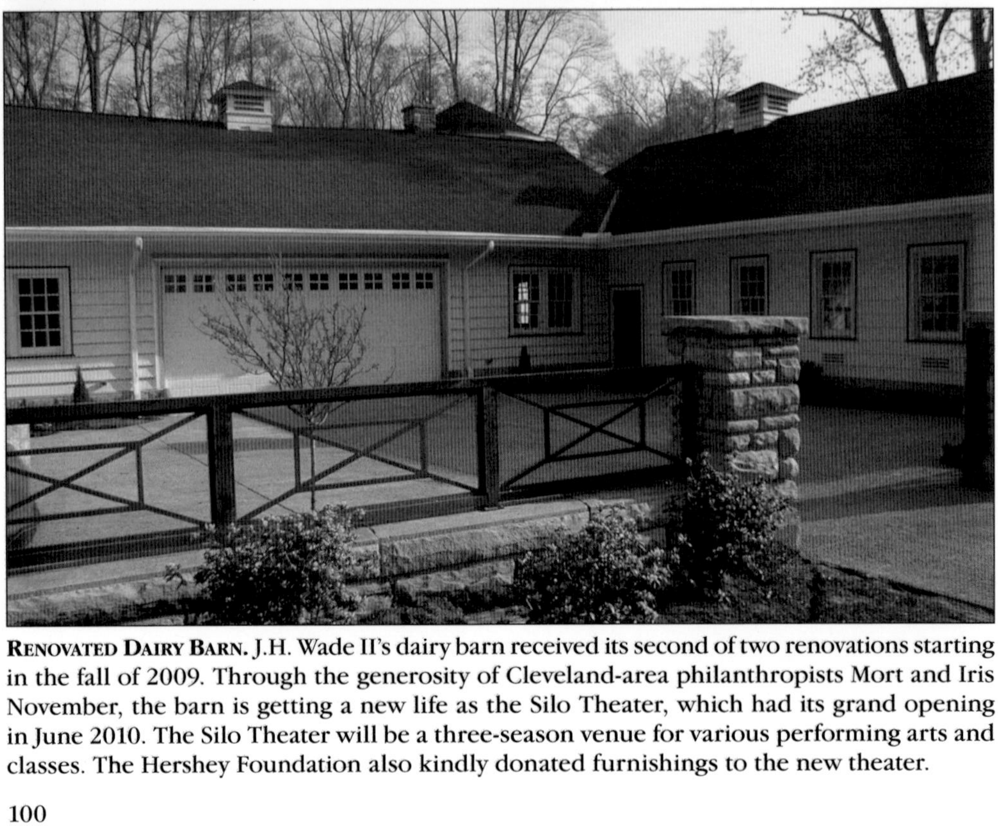

Renovated Dairy Barn. J.H. Wade II's dairy barn received its second of two renovations starting in the fall of 2009. Through the generosity of Cleveland-area philanthropists Mort and Iris November, the barn is getting a new life as the Silo Theater, which had its grand opening in June 2010. The Silo Theater will be a three-season venue for various performing arts and classes. The Hershey Foundation also kindly donated furnishings to the new theater.

Six

University Farm Today
Current Endeavors

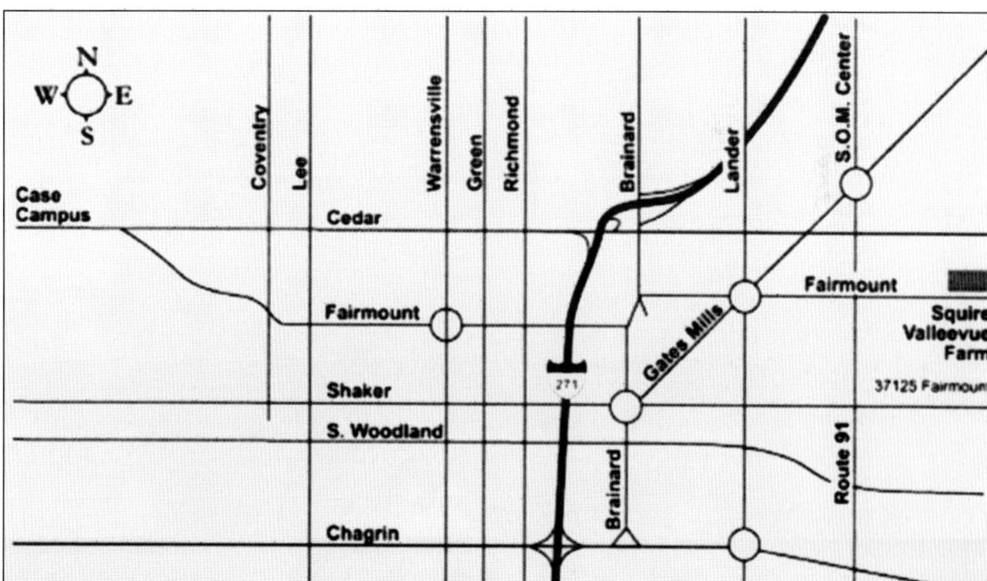

Location of University Farm in Relation to the Main Campus. The farm is an integral part of Case Western Reserve University even though it is physically apart from the main campus. Located 10 miles east in the more rural and wooded portion of Cuyahoga County, it has served for decades as a field research site to complement many of the university's research and educational activities.

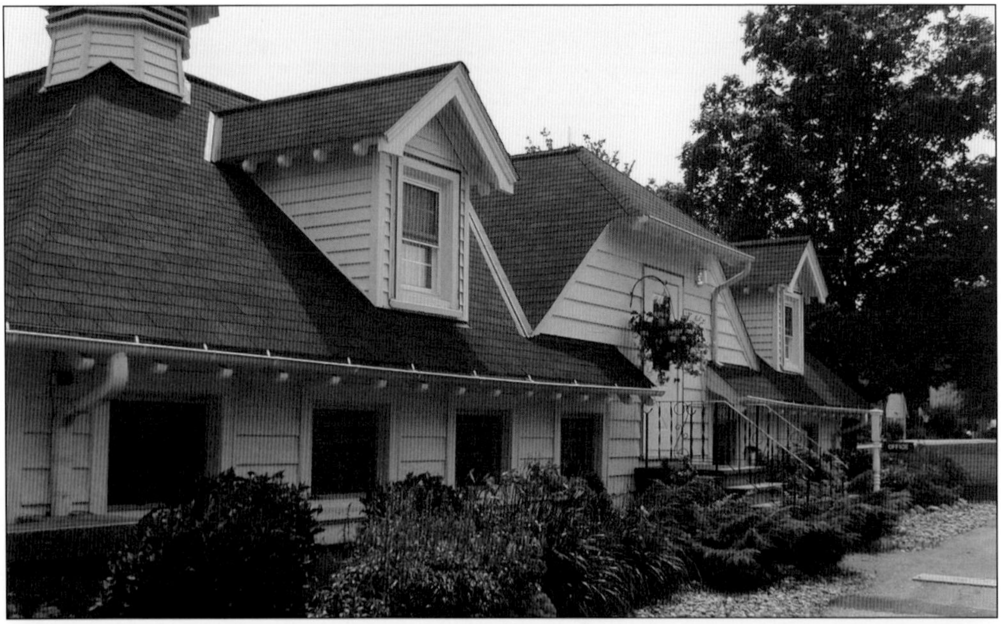

The Main Barn Complex. First built as stables for Andrew Squire's livestock, then converted into a multiuse building including a theater, then renovated into an animal studies facility, the main barn now houses the administrative offices of the farm, as well as classrooms and labs. The Kutina Classroom, Foreman Lab, and the Mather Teaching Lab are the most used facilities within the main barn complex.

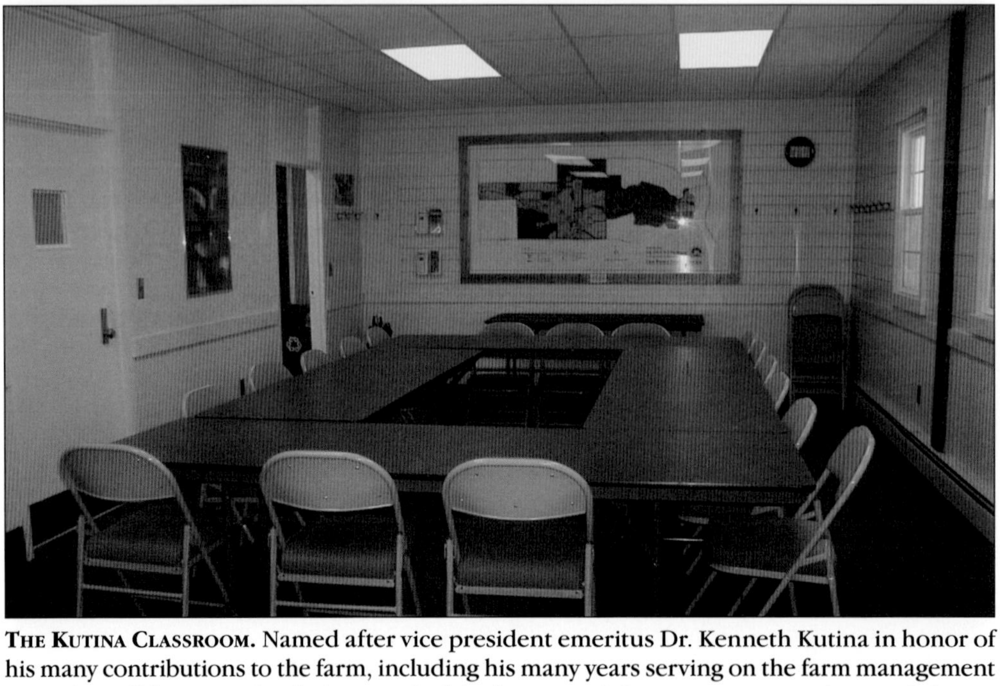

The Kutina Classroom. Named after vice president emeritus Dr. Kenneth Kutina in honor of his many contributions to the farm, including his many years serving on the farm management committee (11 years as its chairperson), this classroom is host to many meetings and classes year-round. The popular Summer in the Country and Autumn in the Country classes and meetings of the local Audubon society are held here.

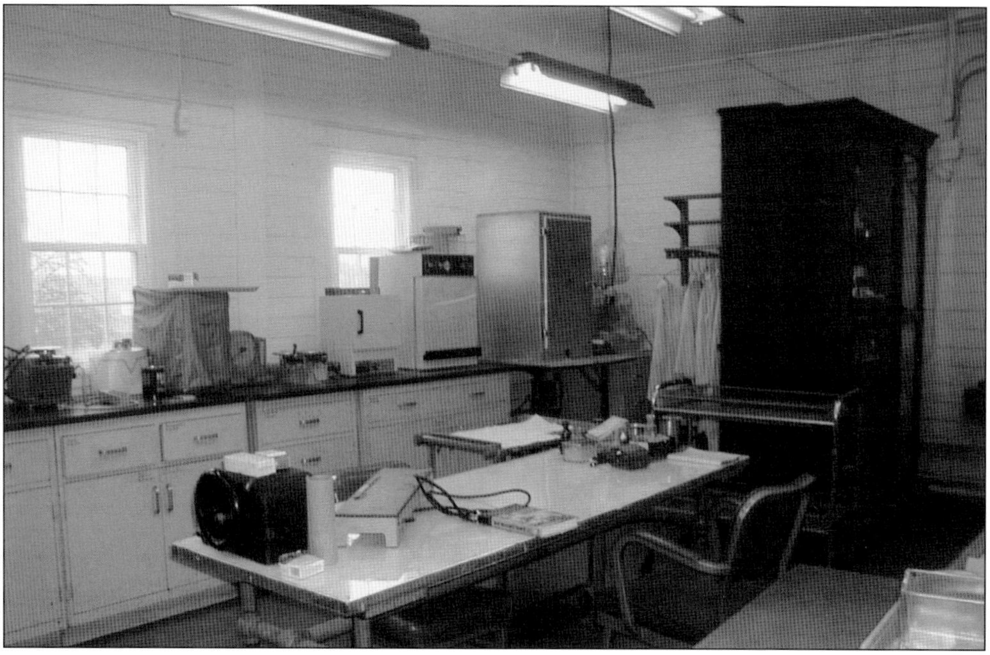

THE FOREMAN LAB. Named after professor emerita Dr. Darhl Foreman, this lab is now used primarily by postgraduate research students as a facility to further their various studies. It was the same lab that Dr. Foreman used for years with her study of prairie dogs. The lab is located directly across the hall from her office.

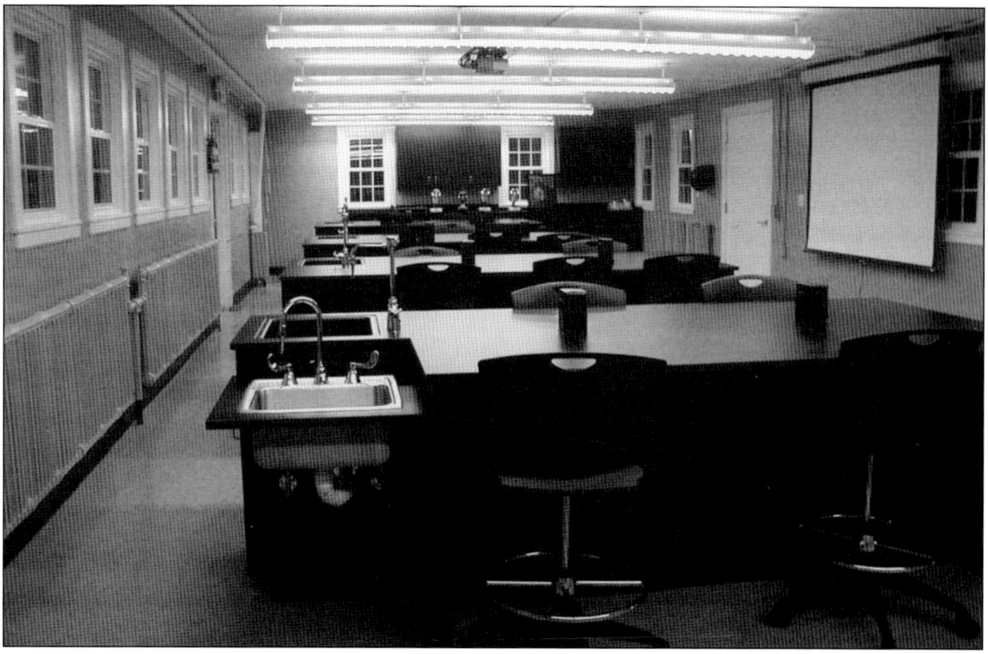

THE MATHER TEACHING LAB. Several different campus disciplines employ this multiuse lab named in honor of Flora Stone Mather. Often, field samples obtained from the research areas of the farm are brought back here for further analysis and study. This lab was updated in 2008 with funds donated by the Flora Stone Mather Alumnae Association.

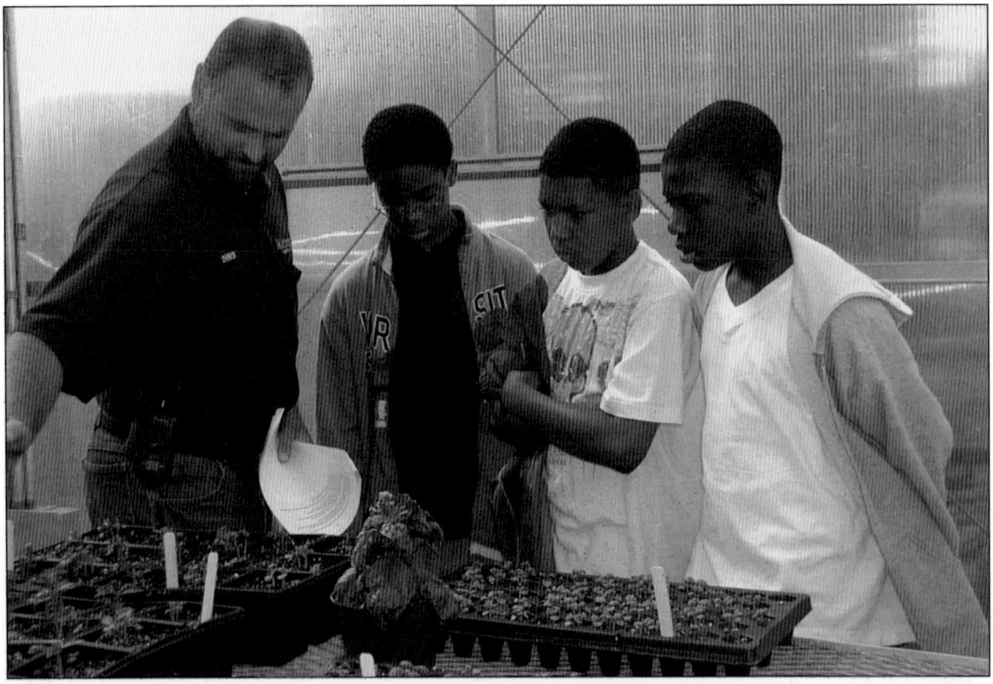

Greenhouse Teaching Facilities. Still in the original 1947 building are two research labs, offices, and a classroom. These facilities are also used by a variety of departments and serve to further geological, ecological, and environmental studies. They are also the site where many of the community education classes are held, such as field trips hosted at the farm for neighboring private and public school groups of all ages. Seen here is farm horticulturist Chris Bond teaching a group of students about different stages of plant growth.

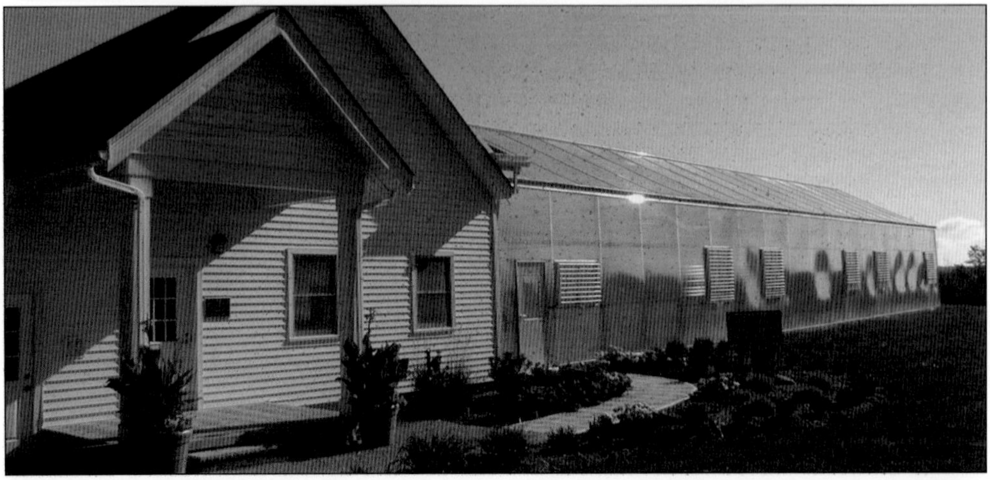

The Debra Ann November Research Greenhouse. With funds generously donated by Morton and Iris November, the greenhouse was renovated in 2006 to replace the original 1947 glass structure. It was named in honor of the Novembers' late daughter, Debra Ann. The greenhouse contains several research rooms that have hosted a variety of studies and disciplines, including genetic manipulation of flax plants by the biology department and carbon sequestration research by the department of civil engineering. Its current primary use is to grow food for the university's farm food program.

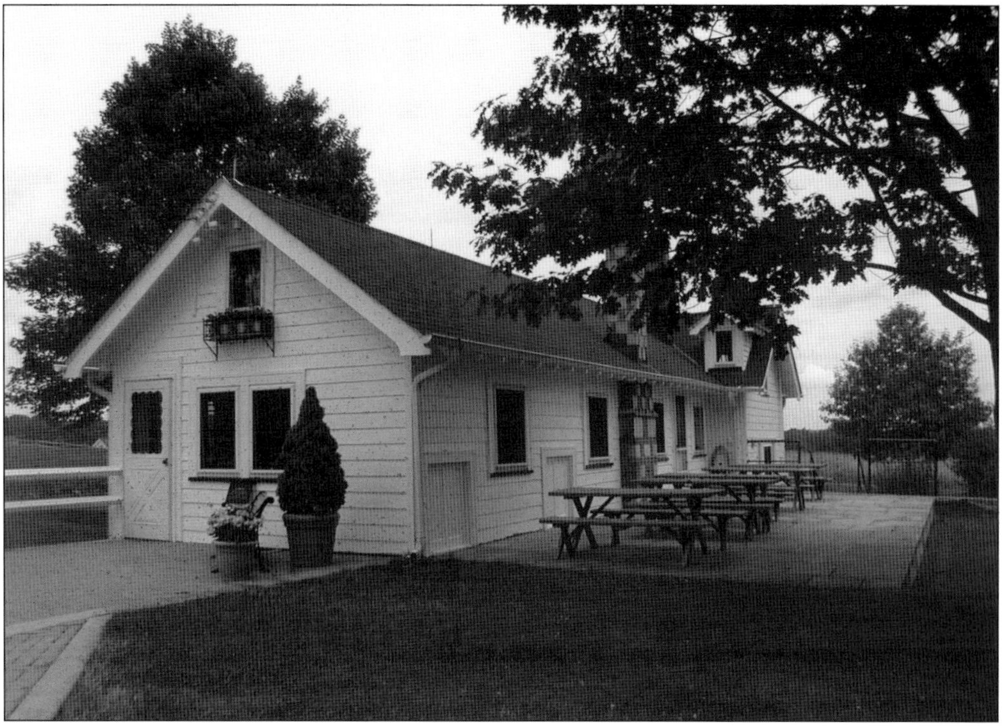

The Pink Pig. Still a popular facility for student groups and department meetings, this quaint cottage accommodates overnight excursions to the farm, just as it did in the days of the Flora Stone Mather College. Students studying various languages will often use the facility for a weekend totally immersed in the language they are learning, with no English allowed.

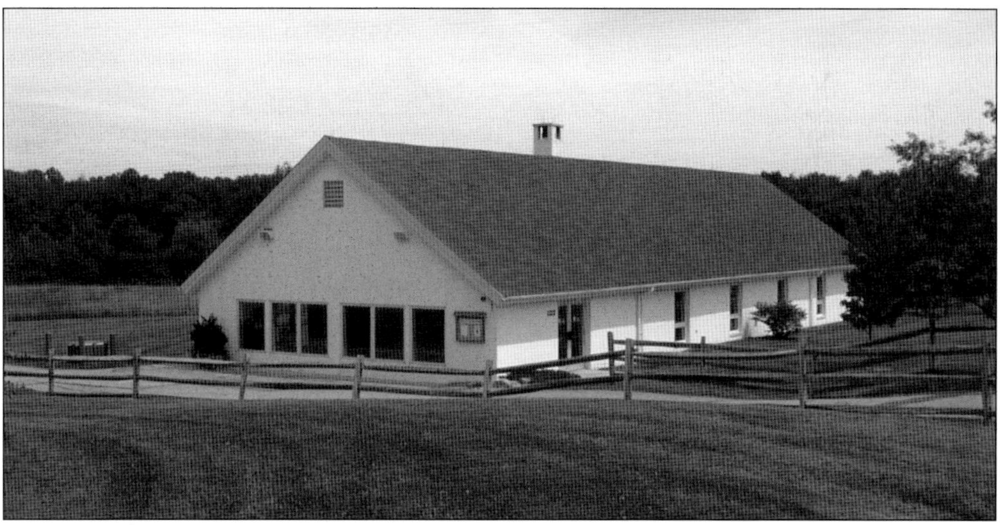

The Sheep Barn. The original building burned to the ground in the 1960s, killing all of the animals inside. This structure was built as its replacement and would house the farm's ungulates for veterinary studies and testing. It was renovated in 1991 and is used as a large gathering place for student groups and department retreats. This renovation effort was made possible by Flora Stone Mather College alumna and former dean, university vice president, and university marshal Patricia Kilpatrick.

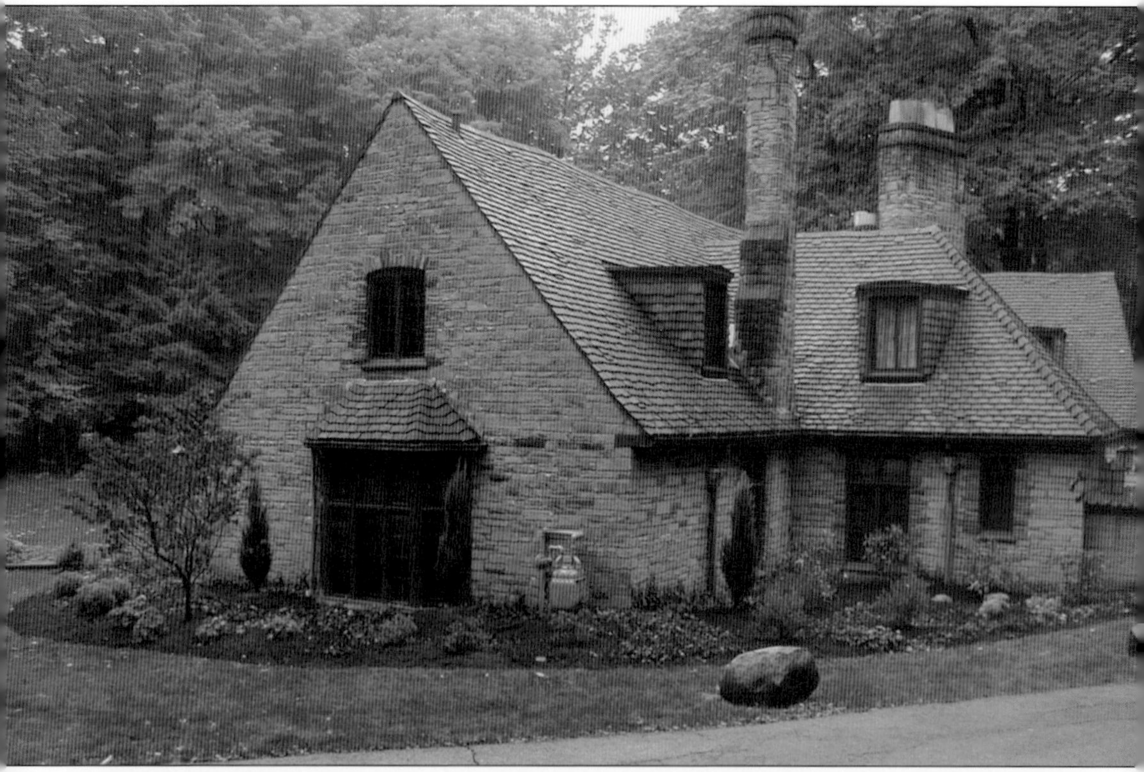

The Manor House Today. The Manor House is still regarded as the most elegant building on the entire farm. After serving as home to both university and farm administrators for several decades, the Manor House now serves as a meeting facility for university departments, recognized groups, and those affiliated with the university. It can be rented by affiliates for private or social events, such as weddings, receptions, and other celebrations.

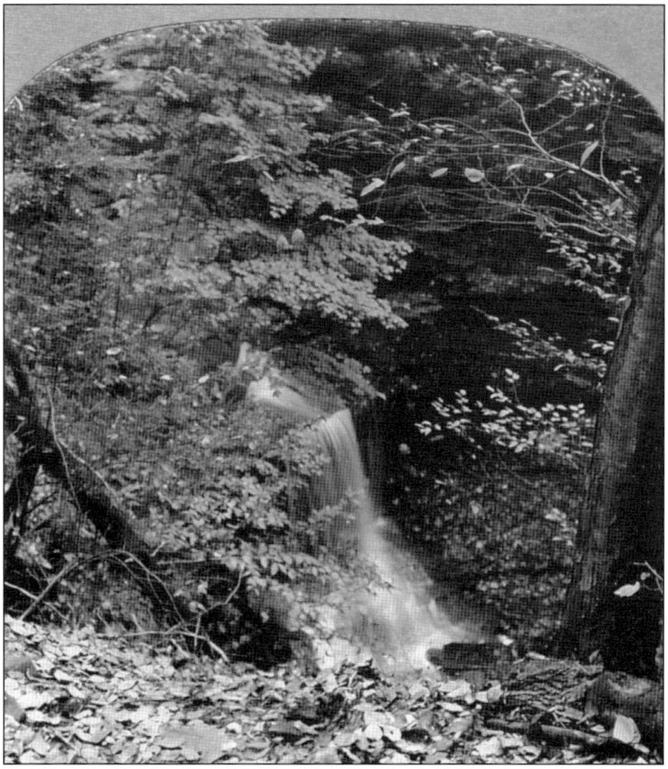

The Waterfall, 1925. One of the hidden treasures at University Farm is this gem, a waterfall that can be seen from one of the farm's many walking trails. Retreating glaciers carved out this beauty, and it roars after heavy rains and snowmelts.

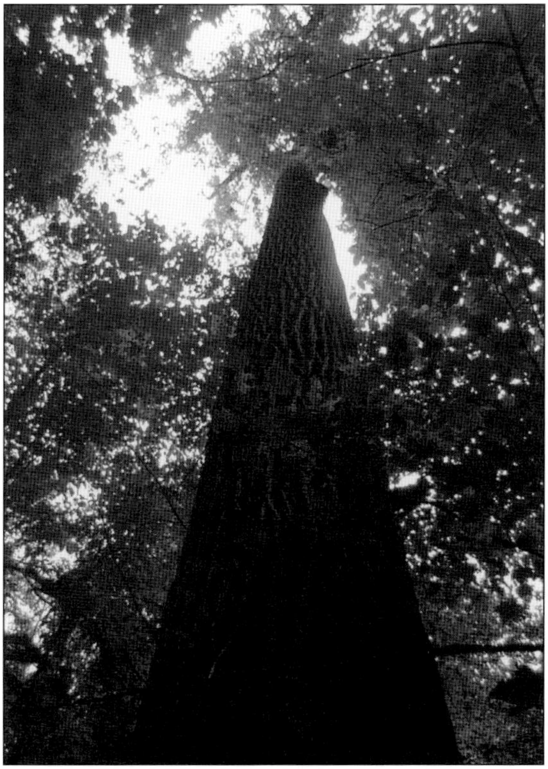

The Woods and Forest. Portions of primary forest that have never been logged are one of the farm's great assets. The primarily maple and beech forest areas are ideal settings for studies conducted by the university and other organizations, such as the Cleveland Museum of Natural History and the Holden Arboretum. Other pristine portions of forest are home to magnificent specimens of hemlock and oak.

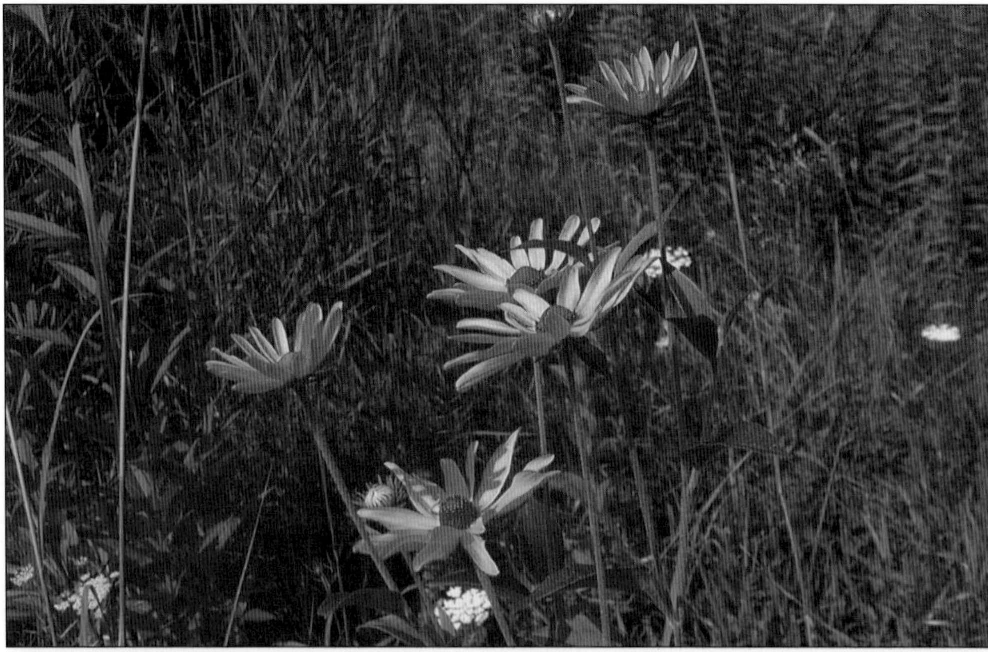

Prairie Restoration. Starting in the spring of 2006 with the collaboration of the Cleveland Museum of Natural History, four acres of old pasture were set aside to facilitate the regrowth of indigenous grasses and forbs species. Every year since, a prescribed early-season burn has taken place to help this restoration process. Funding for the prairie restoration project comes in part from the generosity of Brian and Cindy Murphy.

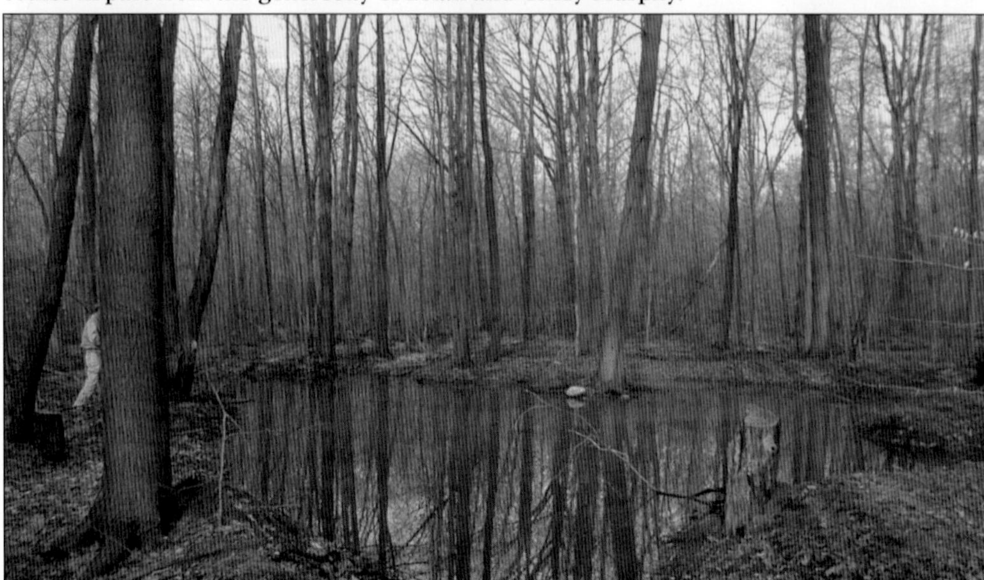

Salamander Lagoon. Installed in 2002, this area of habitat restoration was constructed in the northern portion of the farm property near Cedar Road. Wood frogs and spotted salamanders were introduced to this area to study metal toxicity levels in Northeast Ohio amphibians. Since then, both the diversity and abundance of salamanders and wood frogs has increased at the farm.

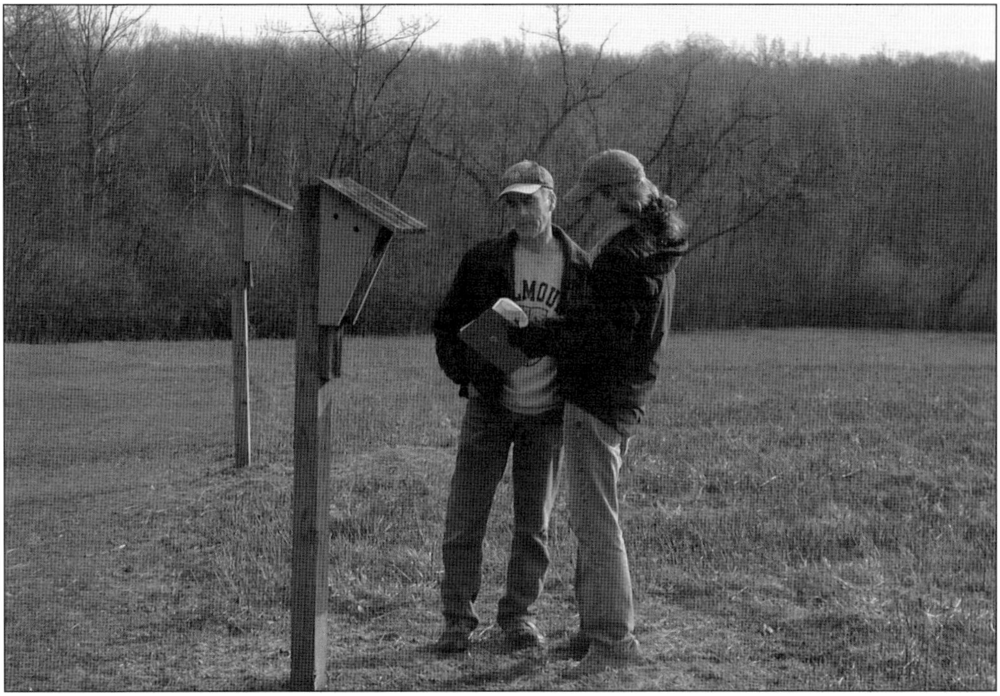

Bluebird Trail. A total of 58 Peterson bluebird houses are situated along portions of the cross-country trail at University Farm. Volunteers, including university alumnus Bill Jirousek and university staff member Betsy Banks, monitor the eggs and population numbers in-season and forward the data to national databases hosted by Cornell University in Ithaca, New York, and the Holden Arboretum in nearby Kirtland, Ohio. Brian and Cindy Murphy have also donated to support this initiative.

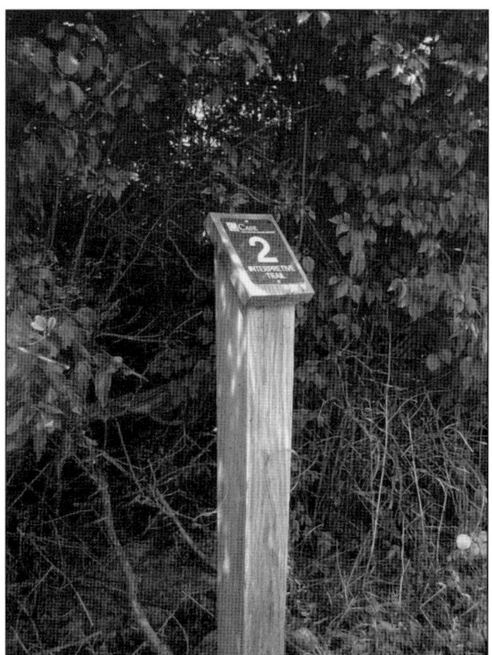

The Interpretive Trail. This two-mile hiking trail designed in 1982 has 32 observation points highlighting wildlife habitats, geological features, and successive stages of vegetation indicative of the farm's diverse features, as well as those that can be found throughout Northeastern Ohio. Along the way, hikers can see old-field communities, climax forest areas, second-growth forest areas, and evidence of many species of wildlife.

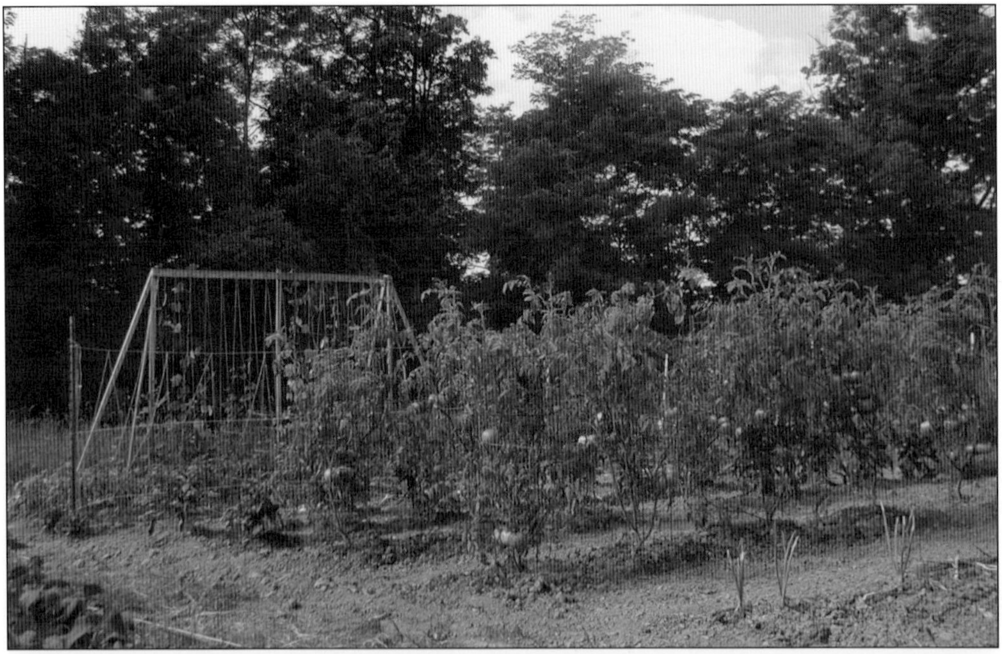

The Community Garden. Available for rent to members of the university community are 21 separate 600-square-foot garden plots that can be used seasonally for those with a bit of farmer in them. Farm staff prepares each plot seasonally by tilling the garden and amending each plot with organic compost.

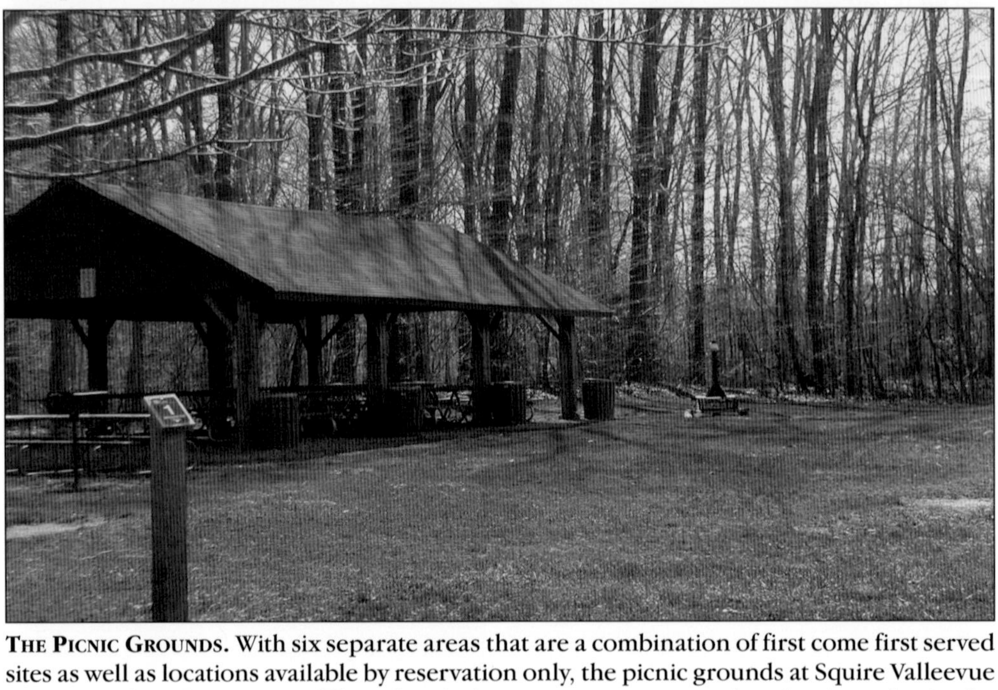

The Picnic Grounds. With six separate areas that are a combination of first come first served sites as well as locations available by reservation only, the picnic grounds at Squire Valleevue Farm is a place for anyone affiliated with the university to unwind and enjoy a day in the country. In addition to volleyball courts and baseball fields, the Flight Nurse Academy's field training for emergency response takes place on the grounds every year in August.

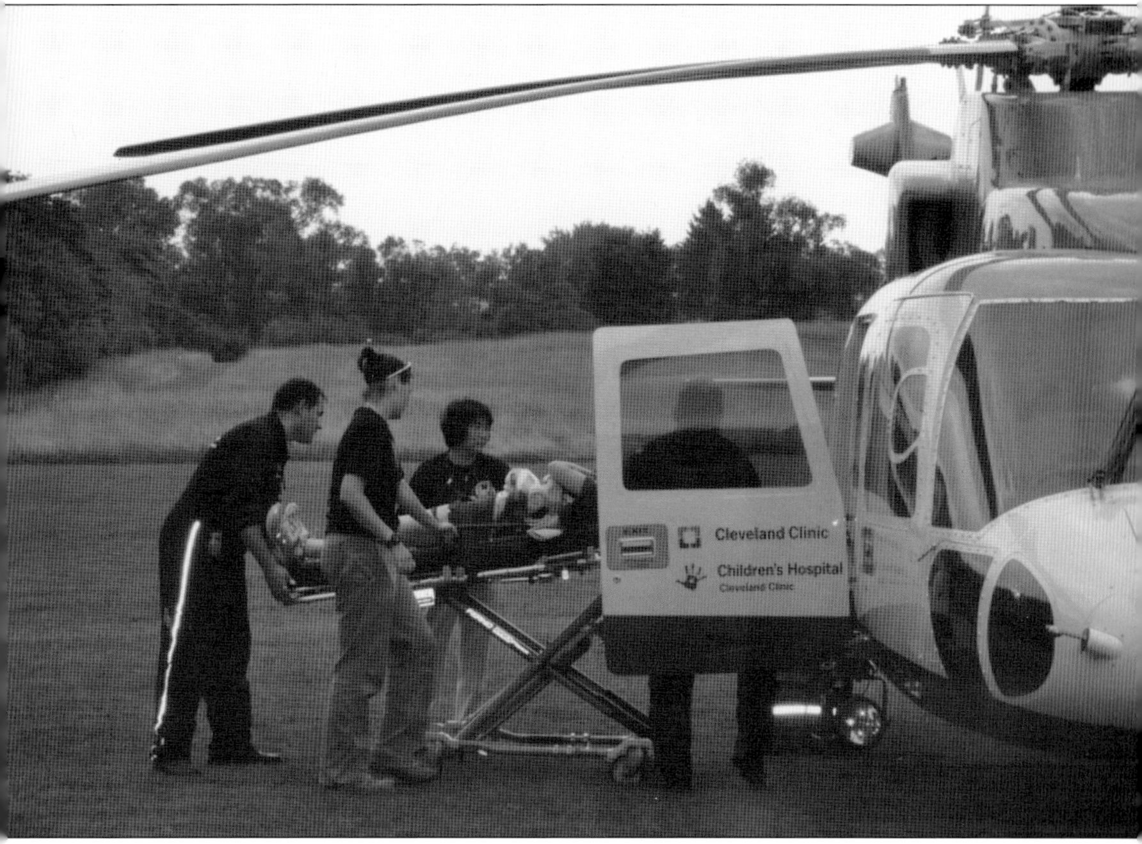

The National Flight Nursing Academy Program. Every August, Case Western Reserve University holds a summer camp for the acute care nurse practitioners, flight nurses, and emergency personnel where a mass-casualty response drill is performed. The camp is the only one of its kind in the country, and it draws students and participants from the entire country and as far away as Japan. The weeklong training course includes pediatric trauma and obstetric emergencies, advanced airway and extrication, central line placement, and suturing labs using high-tech simulated patients. Live helicopters are on hand to aid in flight safety. (Ivan Locci.)

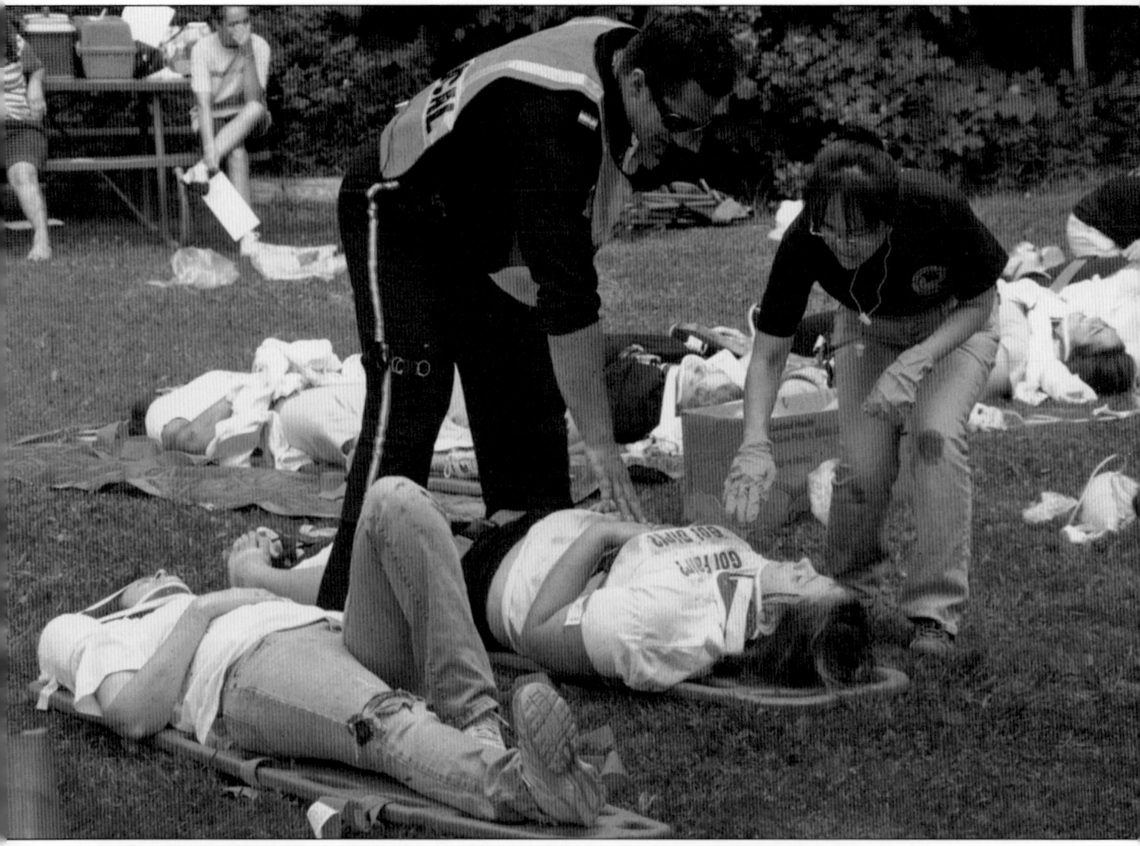

"**Victims**." Hundreds of students and members in the community play the roles of victims in simulated disaster drill scenarios. They are unaware of what role they will be playing in support of the drill. Professional makeup artists spend the entire morning transforming the group into a haggard, blood-covered, shrapnel-filled, and broken-boned mess for the nursing students to sort out and practice triage on. (Ivan Locci.)

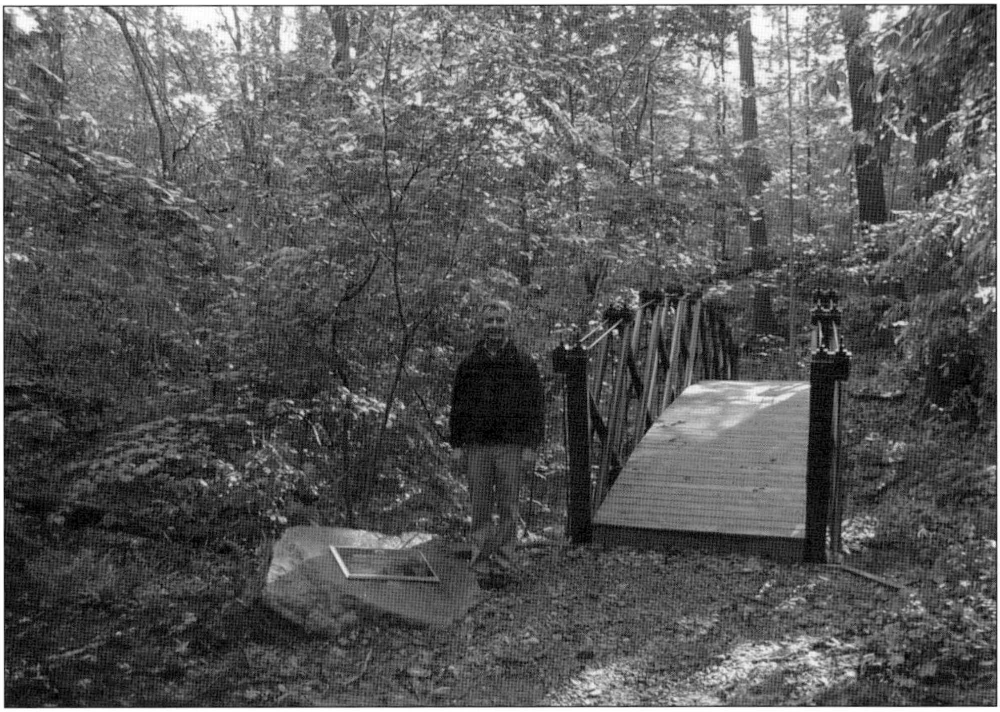

Pedestrian Bridge. Designed and built in 2002 by Prof. Dario Gasparini (pictured) and his engineering students, this pedestrian bridge spans a portion of the creek that was previously difficult to traverse by foot. University Farm offers numerous opportunities for real-world applications of classroom instruction and research.

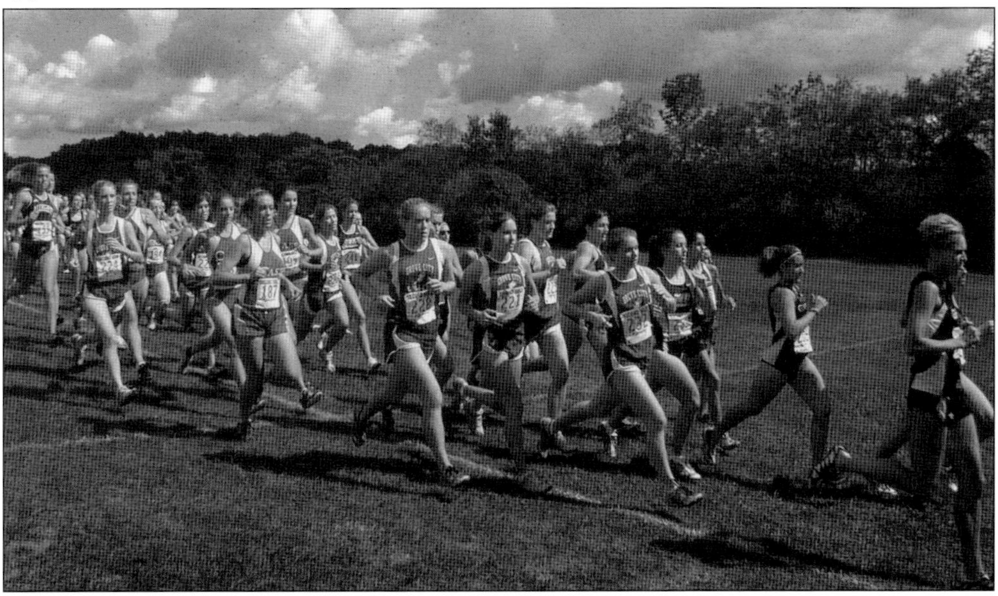

Athletics. In addition to the farm's numerous educational, research, and recreational purposes, it also is used by the Case Western Reserve University's cross-country team for practices and meets. Every year, several races are held at the farm. Area schools also use the five-mile trail for their cross-country team's practices when not in use by the university.

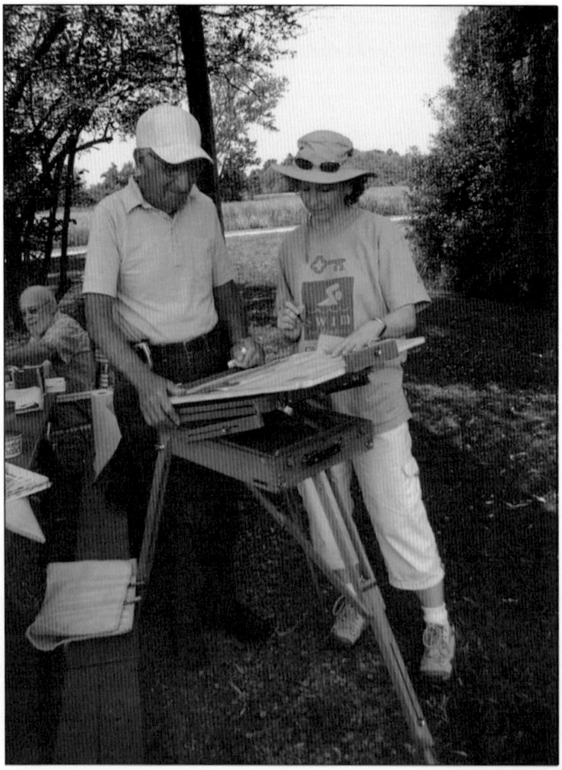

Art Students at University Farm. University Farm currently hosts several art classes, both at the collegiate for-credit level and also recreationally during the Summer in the Country series. Here, some Summer in the Country students and their instructor, Patricia Sigmier, are seen setting up to do alfresco painting in one of the farm's picnic areas.

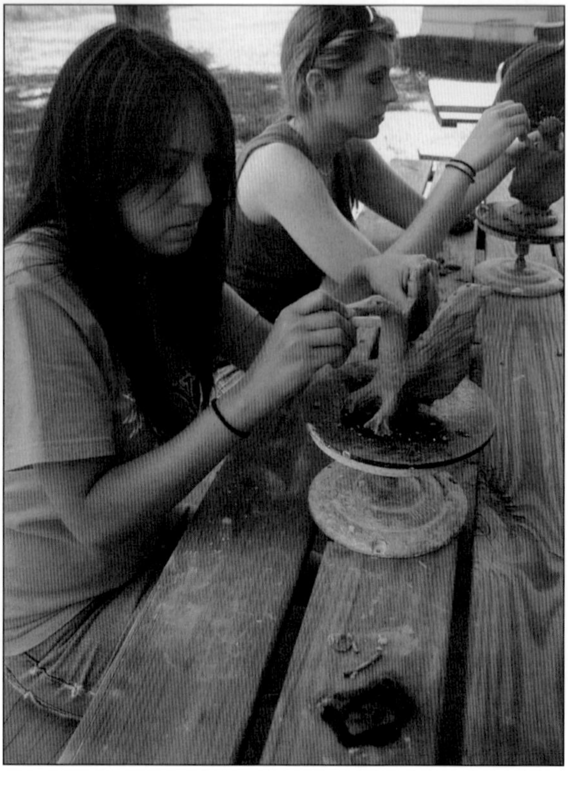

Ceramics Studio. Several university art classes are held at Valley Ridge Farm every year, including traditional ceramics classes and raku ceramics classes, which are held in the renovated pig barn. The students use both indoor and outdoor firing methods for their sculptures and take their inspiration for their work from the nature around them. These students are working outside under the renovated water trough.

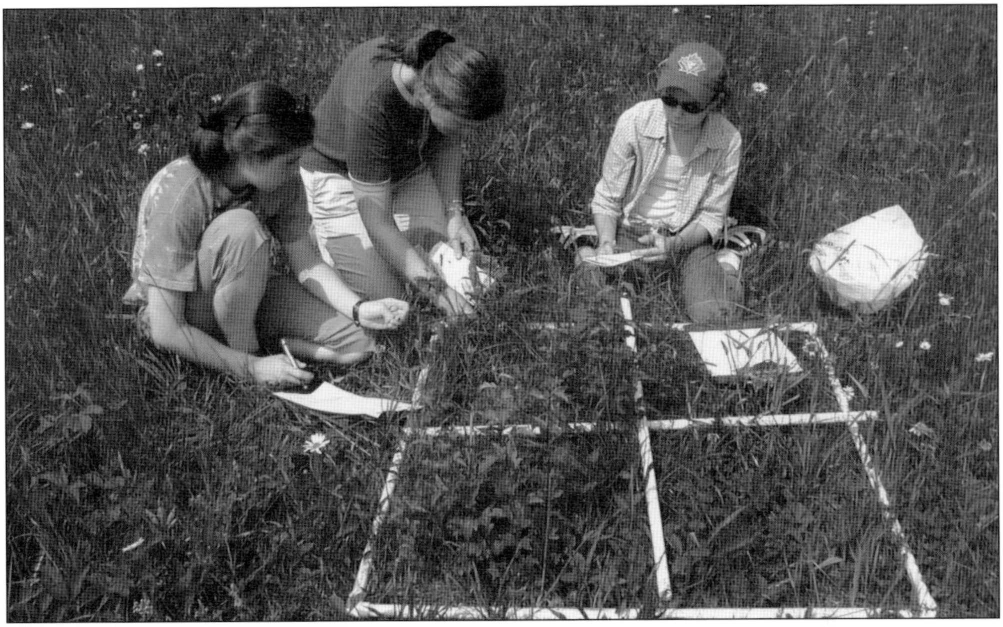

Visiting School Students. Hundreds of area school students are hosted at the farm for field days throughout the year. The farm provides a hands-on environment and experience to supplement their in-class science, biology, and ecology lessons. Pictured are high school students and their instructor, CWRU alumna Stacey Heffernan, using a quadrant to study the diverse grass and forb species on a plot at the farm. The farm also hosts and provides opportunities for many of the Cleveland area's science teachers to get some time in the field.

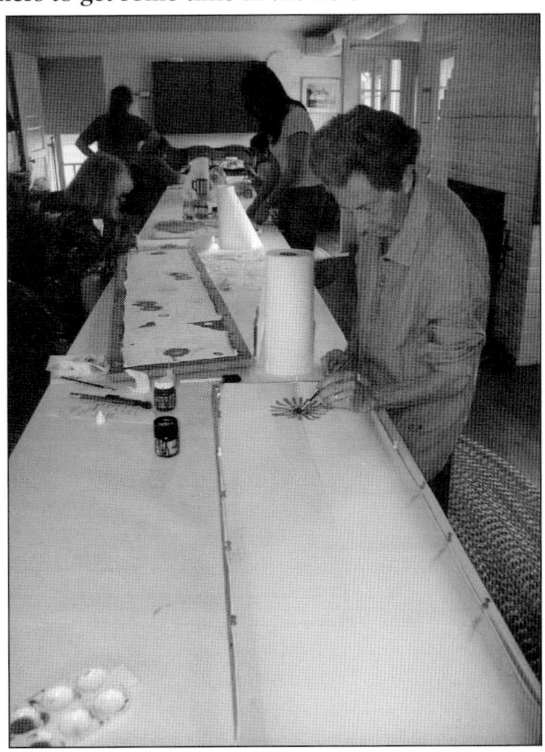

Community Education. Hundreds of the area's community members take advantage of the numerous learning experiences and continuing education offered each season at University Farm. From art and poetry classes to nature hikes and history lessons, the farm truly serves as a campus in the country. Seen here at the Pink Pig is CWRU alumna Mina Kulber in a Summer in the Country class, "Dyes and Silks."

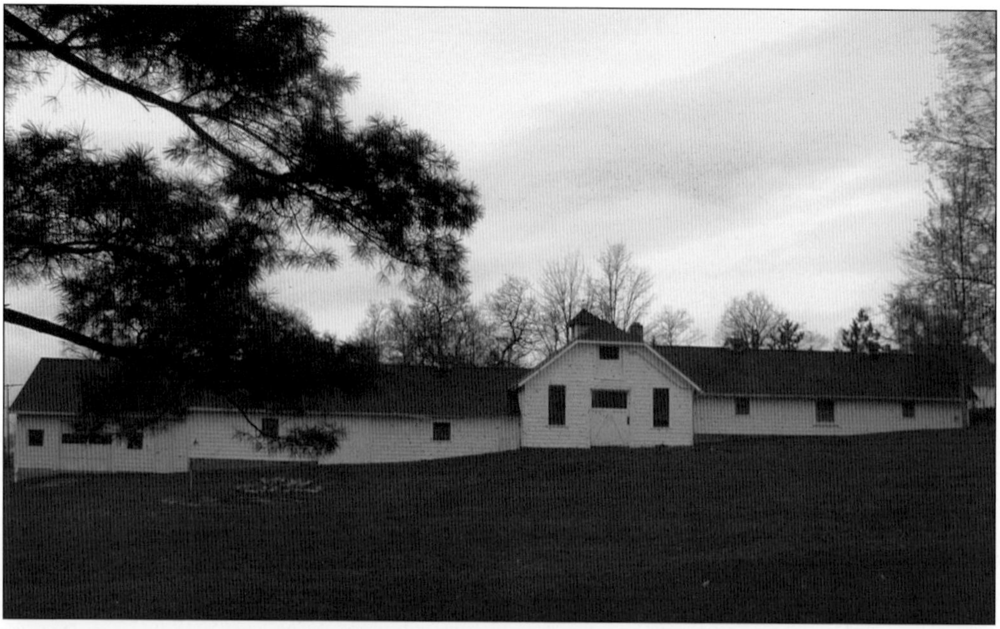

Egg Production. For several decades up until the spring of 2010, University Farm produced over 40,000 fertilized chicken eggs every year to support various research initiatives. They supplied these eggs to labs both on campus and to the Cleveland Clinic. Production was stopped due to increased costs and reduced demand. Pictured is the north side of the Walker and Weeks–designed chicken barn where the eggs were produced for so many years.

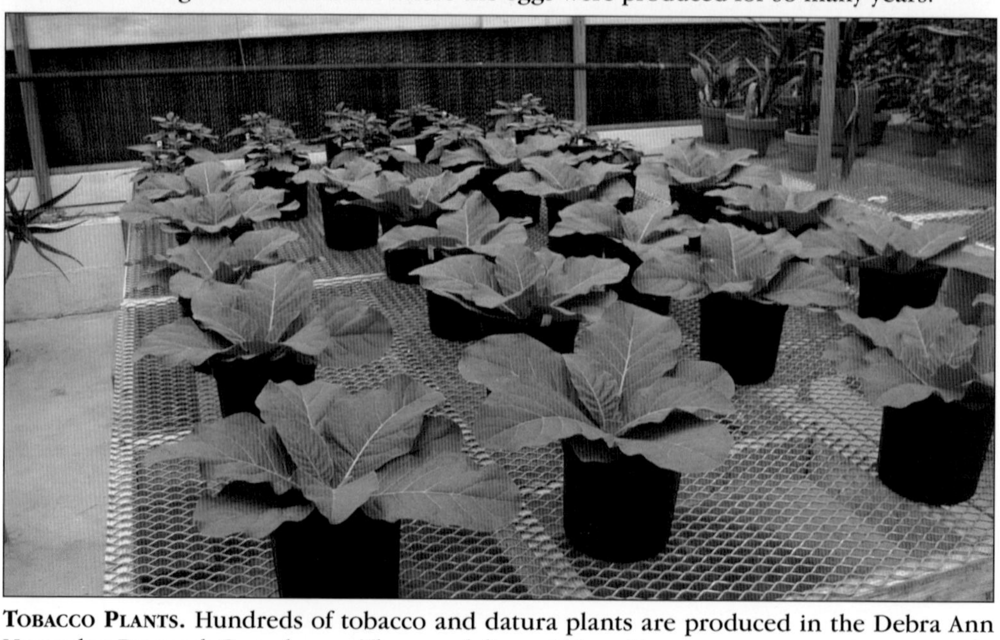

Tobacco Plants. Hundreds of tobacco and datura plants are produced in the Debra Ann November Research Greenhouse. They are delivered biweekly to campus to support research. These plants are not for botanical research as such but serve as the environment in which the tobacco hornworm moth *Manduca sexta* lays its eggs and derives its food source. Associate professor of biology Dr. Mark Willis is studying the sensorimotor and flight patterns of this large insect.

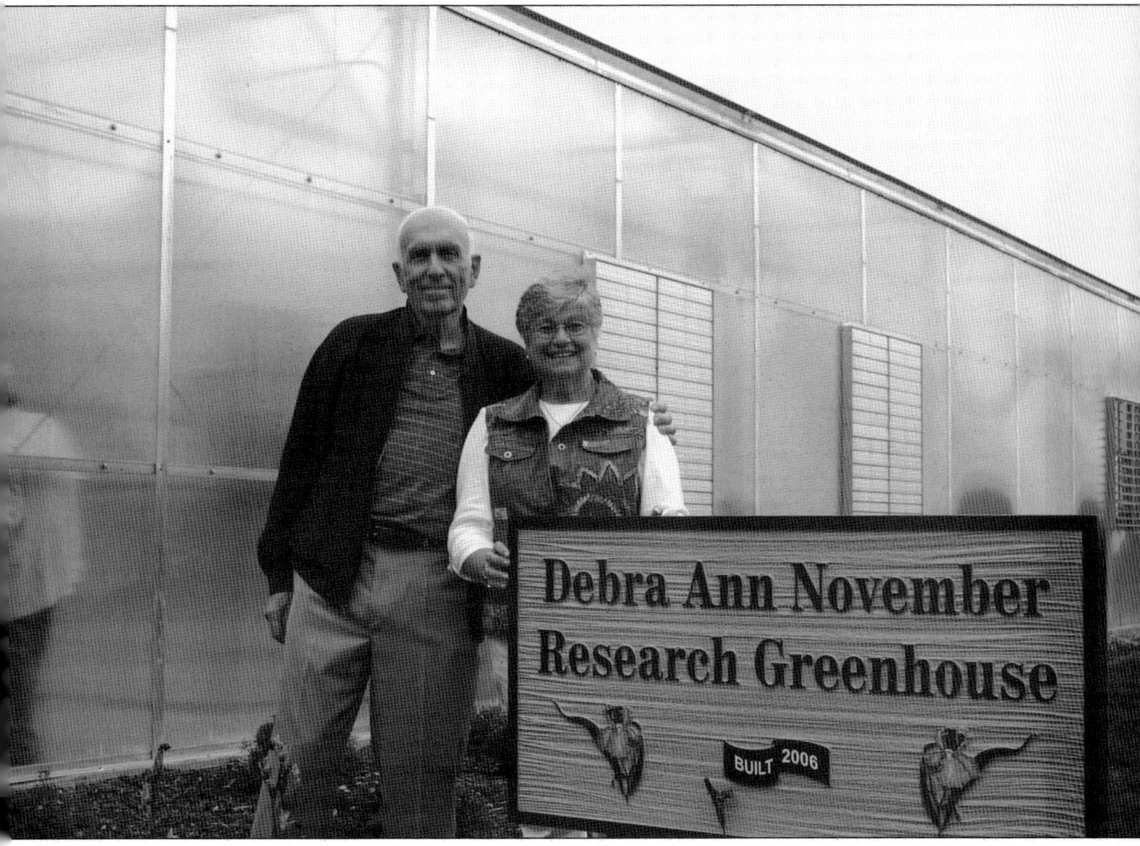

Morton and Iris November. The largest benefactors to University Farm in the 21st century, Morton and Iris November (a Flora Stone Mather College alumna) of Cleveland, Ohio, have generously donated funds. Their donation went to build a new greenhouse, the Debra Ann November Greenhouse, which replaced the original glass greenhouse, and to construct the new theater at Valley Ridge Farm, Silo Theater, which was achieved by converting space in the dairy barn. These two major renovations to the farm could not have occurred without their benevolence.

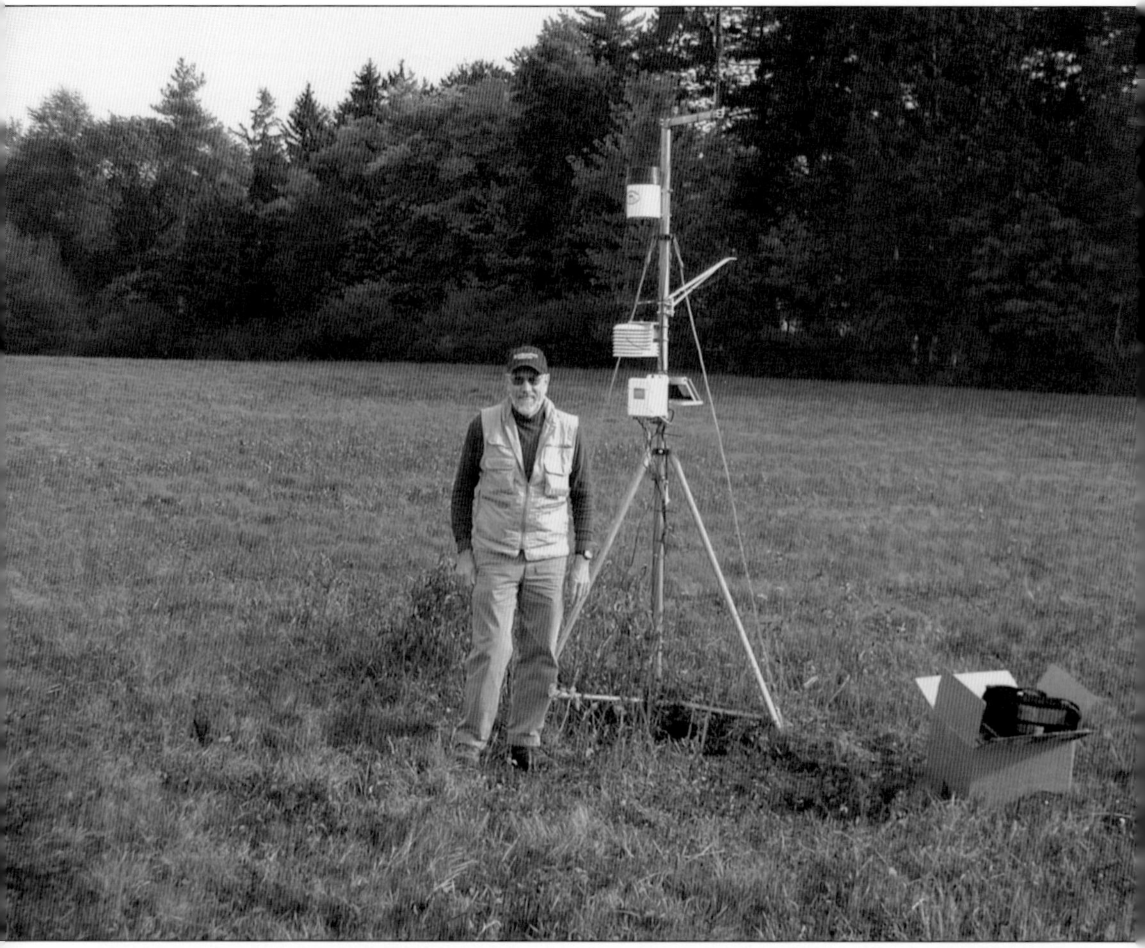

Biology Professor Joseph Koonce. Dr. Koonce is seen here working on a recently installed weather station at the farm. He has served as the chair of the biology department and still serves as a professor. He recently updated these two weather stations at the farm with the goal of establishing a high-resolution environmental monitoring network. These upgraded weather stations take readings every 15 minutes on temperature, humidity, precipitation, and a plethora of other weather phenomena. The readings are automatically logged and archived every 12 hours. This data is available in real time on the farm's website.

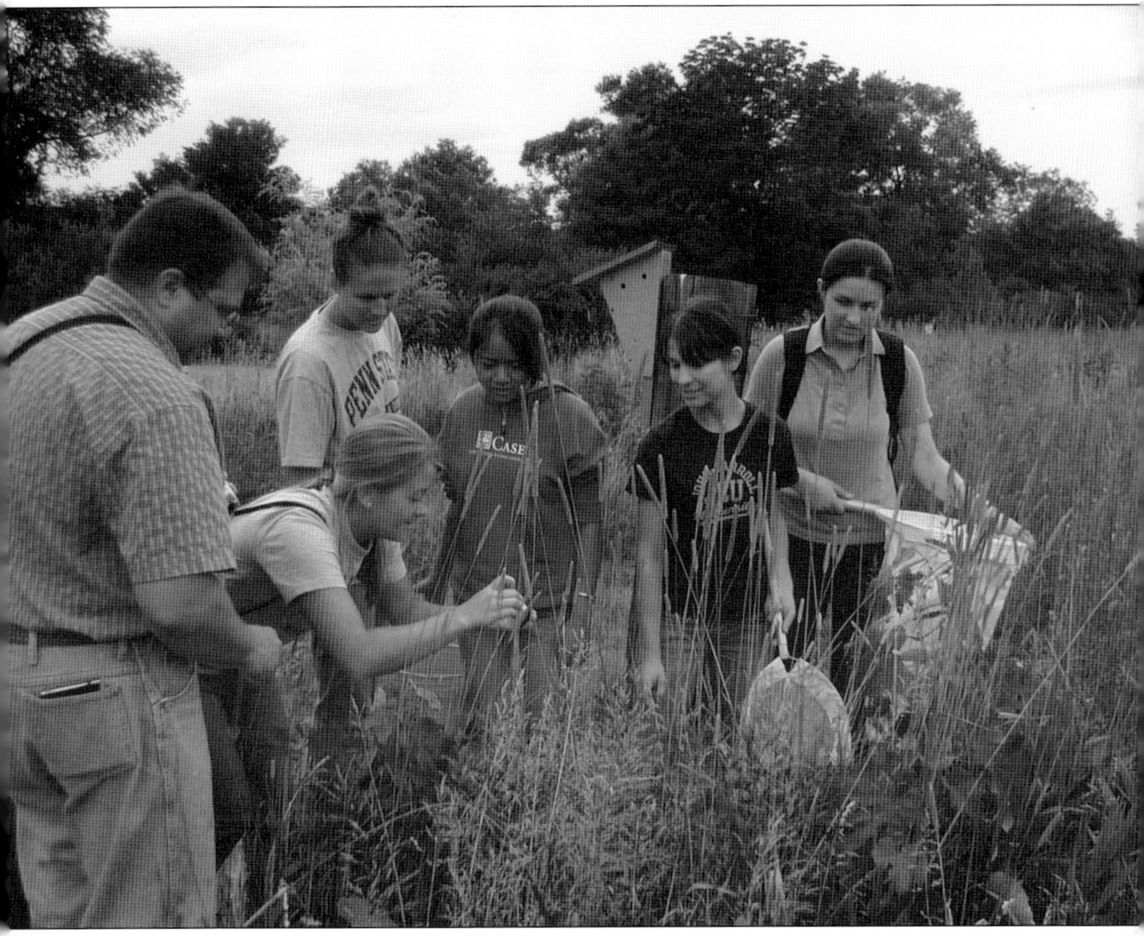

Biology Faculty Jim Bader. Jim Bader (left foreground) is pictured teaching Biology 214 ("Genes and Evolution") to his students on a field trip to University Farm. They collected insects from the fields and analyzed the diversity of species in various microhabitats within the maple-beech forest. Besides teaching biology, Bader is director of the university's Center for Science and Mathematics Education. He is also a Case Western Reserve University alumnus.

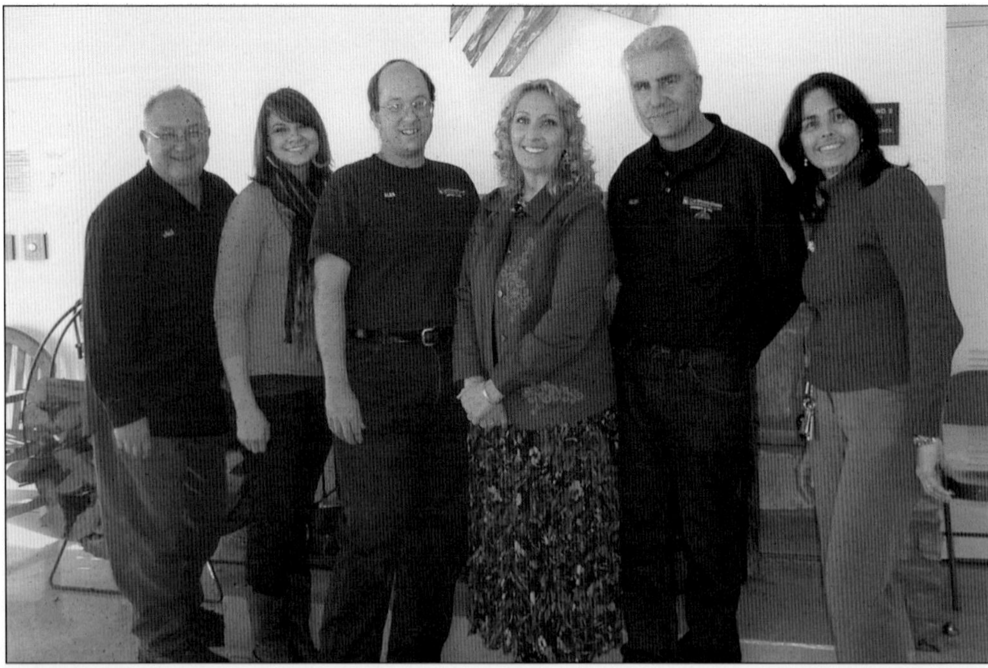

University Farm Staff. By land area, University Farm is by far the largest department in the university, although it is only staffed by a combination of one part-time and six full-time staff members. Included in this image are, from left to right, farm staff John Schwartz, Manor House events coordinator Kimberly Deininger, farm staff Alan Alldridge, department assistant Patty Gregory, farm foreman Mark McGee, and farm director Dr. Ana Locci. Not pictured is farm horticulturist Chris Bond.

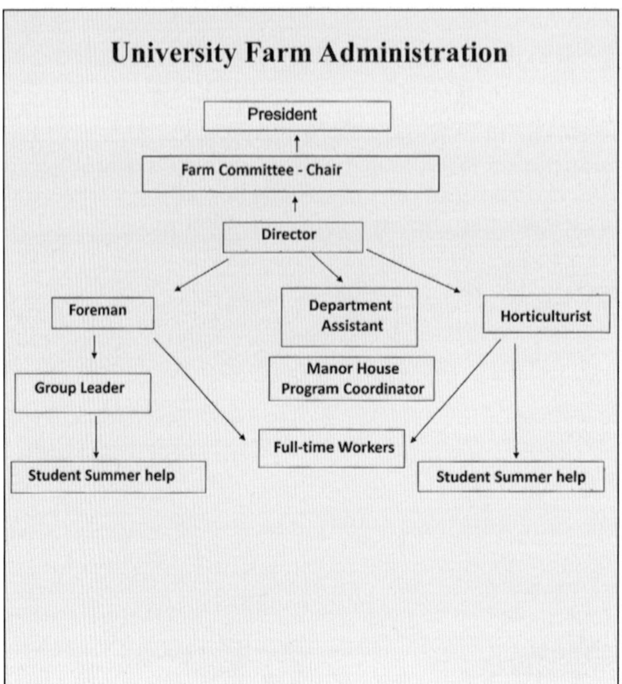

University Farm Administration. Since 1976, the policy and goals of University Farm have been set and overseen by a nine-member panel representing a cross-section of the university. It includes members of the faculty, staff, and administration of both the university as a whole and the farm. The hierarchy of University Farm is seen on this flow chart. The current members of the Farm Management Committee are David Hutter, PhD; Kenneth Kutina, PhD; Ana Locci, PhD; Heidi Martin, PhD; David McCoy, PhD; Glenn Nicholls, committee chairman; Beverly Saylor, PhD; Jerrold Scott, MFA; and David Bell, JD.

Seven

UNIVERSITY FARM
FUTURE ENDEAVORS

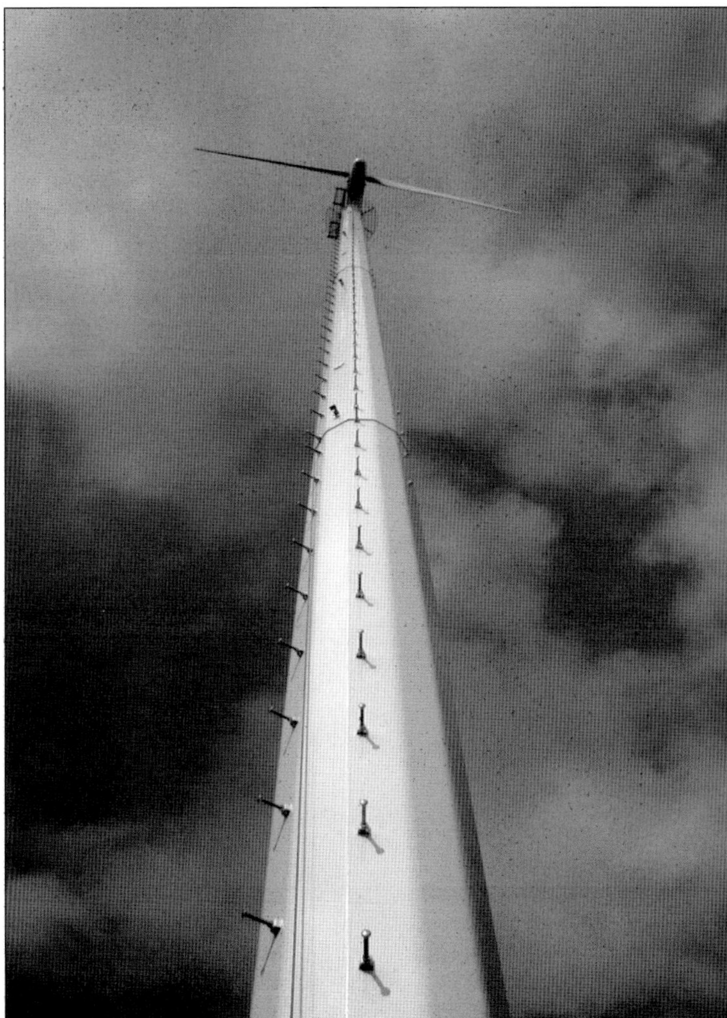

THE FUTURE OF THE FARM. A wind turbine like this is currently proposed for the farm in 2011. Upon installation, it will, coupled with solar panels on the roof of the sheep barn, provide the majority of the farm's energy needs. Sustainable energy initiatives, along with sustainable agriculture practices, are the current goals of University Farm while forging ahead through the 21st century.

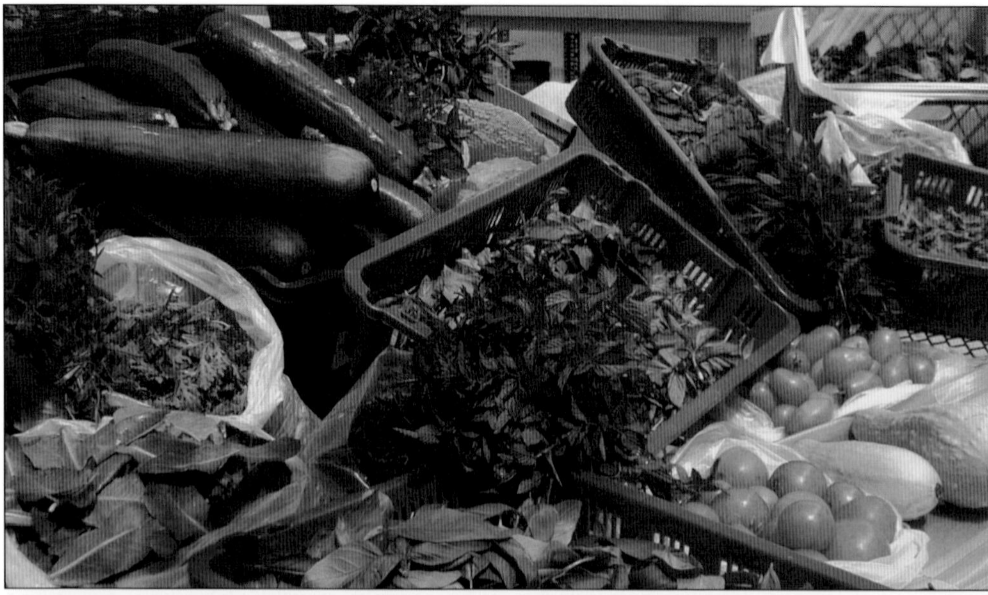

Sustainable Agriculture. Starting in 2010, food grown organically was produced in the on-site greenhouse, the community gardens, Valley Ridge Farm fields, and other smaller on-site gardens. The university's campus dining service, the Bon Appétit Management Company, helped to get the project started with a donation and continues to be the program's strongest supporter. They purchase nearly all of the products produced at the farm and serve them to the students, staff, and faculty.

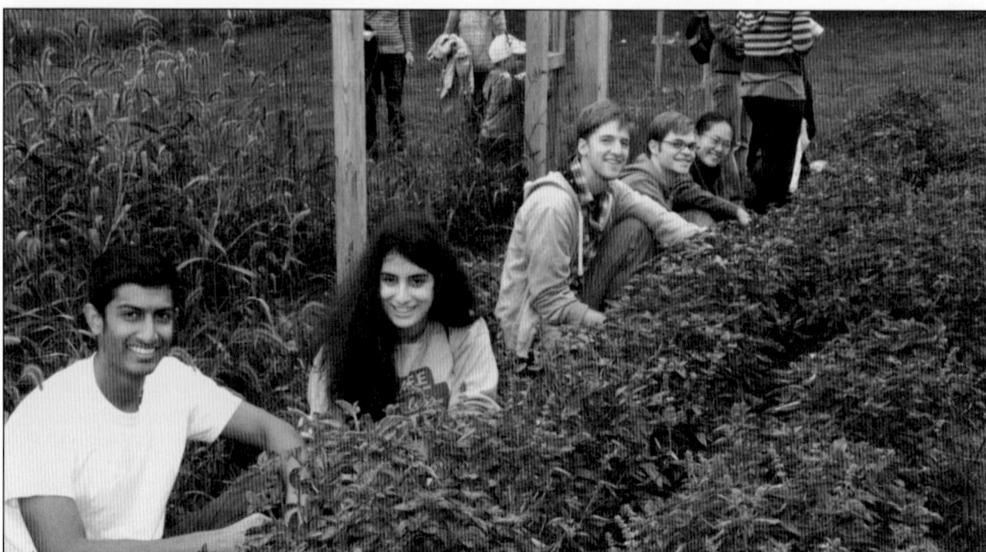

Harvesting. With the assistance of full-time and part-time paid and volunteer student labor, the farm was able to produce over three tons of organically raised food in the inaugural year of the farm food program. This was done on approximately one half acre. This farming endeavor will expand to over two acres in 2011, and there are plans to expand it up to five acres within the next two to three years. The size of the harvest ranged from a low of less than two pounds at the beginning of the year to a high of over 900 pounds in one week at the season's peak.

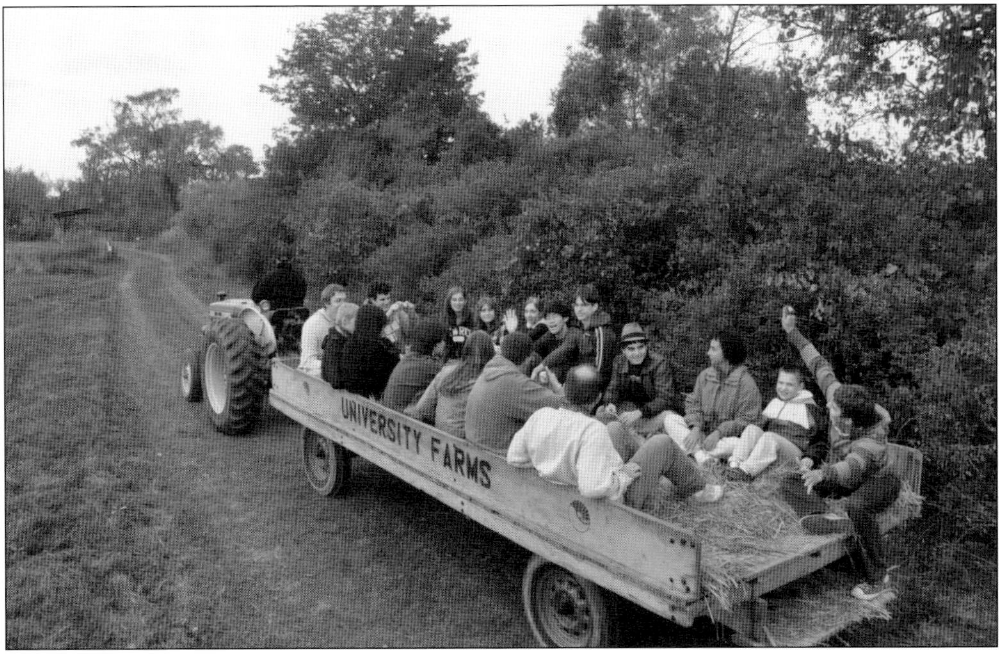

The Fall Harvest Festival. Reviving a tradition from the Flora Stone Mather College era, in 2010, the former Faculty Farm Days were reborn as the Fall Harvest Festival. Organized by Eric Hamilton of the Student Sustainability Council, the Division of Student Affairs, and University Farm staff, this new tradition proved to be a successful and well-attended event.

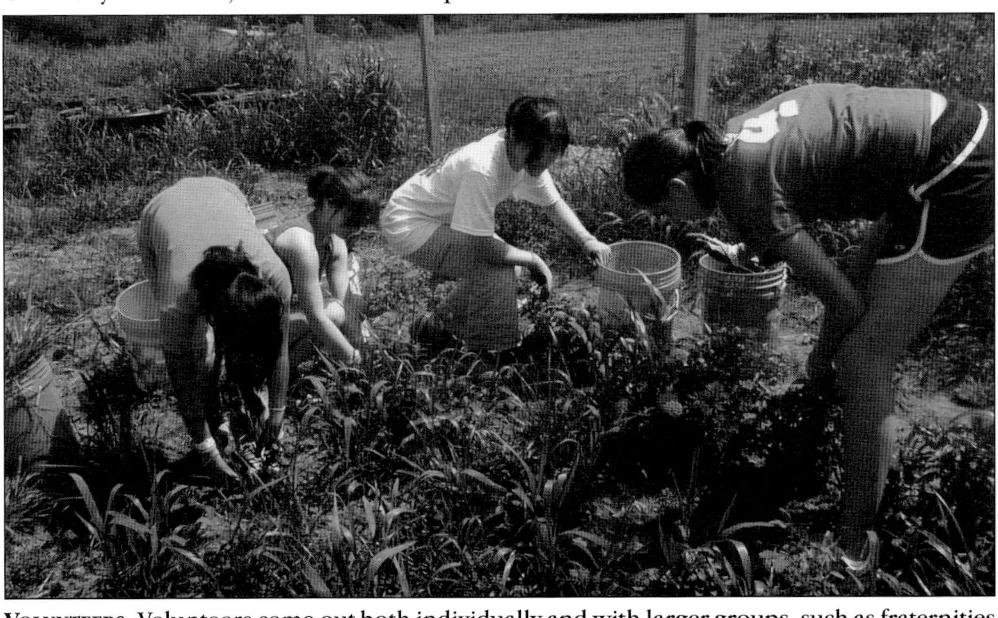

Volunteers. Volunteers come out both individually and with larger groups, such as fraternities and sororities. Though mostly Case Western Reserve University students, volunteers have come from the ranks of university staff, students, and faculty from other schools and members of the community at large. Approximately 20 percent of the hours dedicated to the farm food program have been performed by volunteer labor. University Farm is grateful for services—past, present, and future.

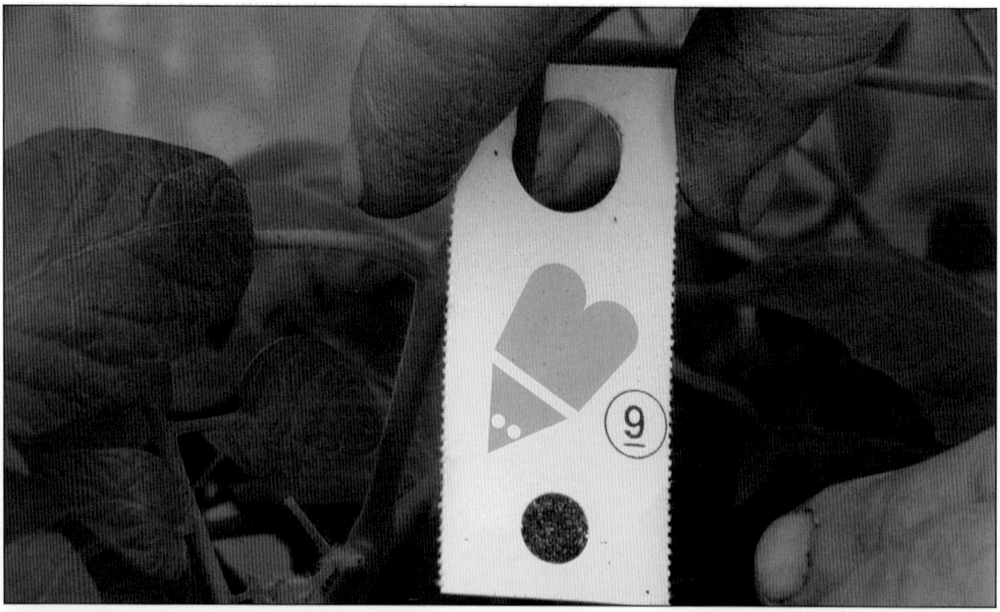

No Chemicals. Environmentally friendly and sustainable methods used to control insect and diseases on plants include the use of natural enemies like lady bugs, beetles, nematodes, and predatory wasps to control insects, and an effort to make growing conditions as favorable as possible to control diseases. Other remedies include the use of plant-derived substances, such as the oil from the neem tree in India. University Farm is committed to raising food for its students without the use of pesticides, herbicides, or any other harmful chemicals.

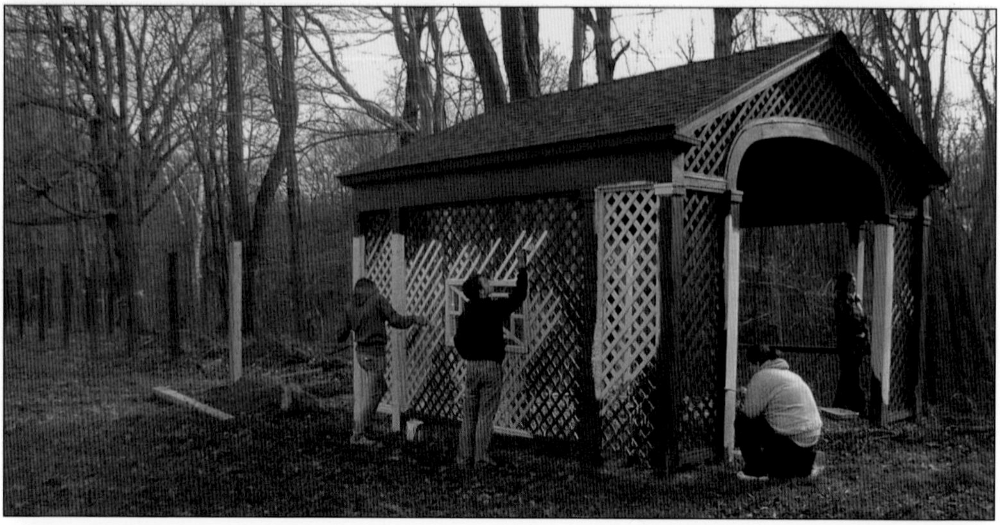

Landscape Restoration. Starting with a section of the original 1910 Warren Manning landscape drawing, restoration efforts at Case Western Reserve University's Valley Ridge Farm began in 2009 to preserve what is left of Jeptha Homer Wade II's original picturesque landscape through the generous donation of Ray and Mary Ann Leonard Zukowski. These restoration efforts will be ongoing, as funding and resources continue to become available to assist University Farm in these endeavors. Should additional funding ever become available, a restoration of what is left of the original Squire Arboretum at Squire Valleevue Farm also will be undertaken.

The Farm. There was no definitive history of Case Western Reserve University Farm. Throughout the years and in the farm's many incarnations, there have been some articles written about it and certain aspects of its buildings and programs, which have been documented for posterity. The authors wanted to create a more complete history of the Farm that ties all of its many stories together and to set the stage for whatever its future history may be.

BIBLIOGRAPHY

Alderson, Ernestine. "A Walk . . . Through the Woods With Andrew Squire." *Your Garden and Home* (January 1931):13–15.

Andrews, Emily R. "The Pink Pig—An Adventure in Outing." *Journal of the American Association for Health, Physical Education, and Recreation* 62 (January 1949): 12–14.

Hively, Suzanne. "Farm Once Grew Medicine at CWRU's Magic Drugstore." The *Cleveland Plain Dealer*.

Kirkwood, Marie. "Squire Valleevue Farm Provides Students Unusual Recreational and Laboratory Facilities." The *Voice of Reserve* (May 1942):3,5.

Kohn, John. "History of Squire Valleevue Farm." Unpublished manuscript, 1981.

Tittle, Diana, and Mark Gottlieb. *Hunting Valley: A History*. Cleveland, OH: Hunting Valley Historical Society, 1999.

About University Farm

Case Western Reserve University Farm, which is comprised of Squire Valleevue and Valley Ridge Farms, is a beautiful 389-acre property that features forests, ravines, waterfalls, meadows, ponds, and a self-contained natural watershed. The farm is the site of several buildings designed for educational, research, and recreational purposes. It is open and available to all students, faculty, staff, and alumni.

The goal of the University Farm is to enhance and complement the university's mission of education, research, and community service through its unique natural settings, history, and cultural features. The farm also supports and enhances additional university community activities, including athletics and recreational events.

As we move forward, University Farm, along with the university's Sustainability Alliance and the student-run Student Sustainability Council, will continue to lead the way in research and implementation of more sustainable methods of energy and food production.

We will continue to be responsible stewards of the land and guardians of the rich histories of the properties. We will be leaders, teachers, and researchers to further the collective goals of the original owners, Andrew Squire and Jeptha Homer Wade II, and the goals and needs of our thousands of students, faculty, and staff—past, present, and future.

To learn more about University Farm and our continued commitments to these goals, and to learn about current goings-on, please visit our website at studentaffairs.case.edu/farm/.

To learn more about Case Western Reserve University's overall commitment to sustainability, please visit its website: www.case.edu/sustainability.

To learn more about University Farm's partner in the Farm Food Program, the Bon Appétit Management Company, and their commitment to sustainable agriculture, please visit www.bamco.com.

Find *Your* Place in History.

www.arcadiapublishing.com

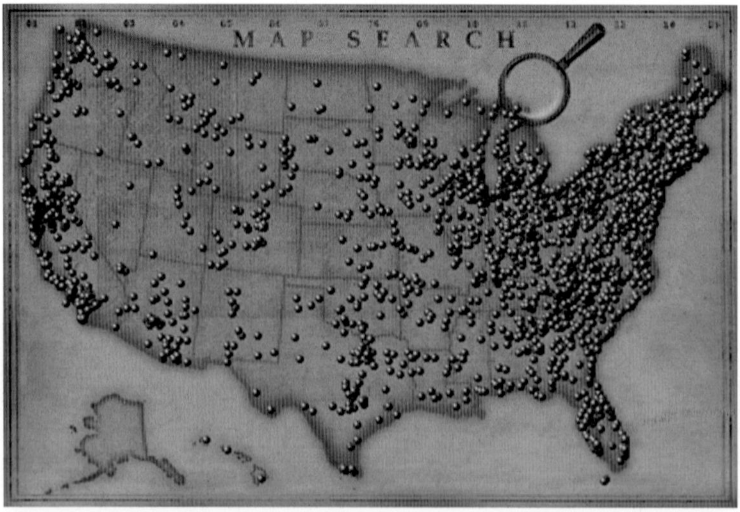

Discover books about the town where you grew up, the cities where your friends and families live, the town where your parents met, or even that retirement spot you've been dreaming about. Our Web site provides history lovers with exclusive deals, advanced notification about new titles, e-mail alerts of author events, and much more.

MADE IN THE USA

Arcadia Publishing, the leading local history publisher in the United States, is committed to making history accessible and meaningful through publishing books that celebrate and preserve the heritage of America's people and places. Consistent with our mission to preserve history on a local level, this book was printed in South Carolina on American-made paper and manufactured entirely in the United States.

This book carries the accredited Forest Stewardship Council (FSC) label and is printed on 100 percent FSC-certified paper. Products carrying the FSC label are independently certified to assure consumers that they come from forests that are managed to meet the social, economic, and ecological needs of present and future generations.

FSC
Mixed Sources
Product group from well-managed forests and other controlled sources
Cert no. SW-COC-001530
www.fsc.org